Hip Arthrography

Hip Arthrography

Paul Grech

M.D., F.R.C.R., D.M.R.D., D.T.M.&H.

*Consultant Radiologist, Northern General
Hospital and Lodge Moor Hospital, Sheffield;
Honorary Clinical Lecturer in Radiodiagnosis,
University of Sheffield.*

CHAPMAN & HALL
London

LON Q ML FFH

First published 1977
by Chapman and Hall Ltd,
11 New Fetter Lane, London, EC4P 4EE

© 1977 Paul Grech

Printed in Great Britain by
T. & A. Constable Ltd, Edinburgh

ISBN 0 412 15090 5

Contents

Foreword

Arthrography of the hip has been practised for over 40 years, but its value as a useful method of investigation has not been generally accepted. Difficulties in technique and interpretation may explain some lack of appreciation.

Arthrography of any joint should be practised regularly to facilitate technique and to gain the necessary experience of interpretation. The surgeon or radiologist interested in this technique should have access to a regular flow of patients.

The arthrogram can produce valuable information regarding the pathology of the condition which may influence treatment or prove of interest in retrospective study. In congenital dislocation of the hip routine arthrography in the past led to greater understanding of intra-articular obstruction and to the appreciation of some complications of treatment in the early moments of life.

Dr Grech has made a stimulating contribution in this monograph which describes technique in the greatest detail. He provides useful guidelines for interpretation and detailed case records which will attract clinical interest. All hip conditions where arthrography might be indicated are described.

The author's work is addressed primarily to radiologists or surgeons who might wish to practise arthrography, but there is much in this monograph which will interest those who are already familiar with the technique.

George P. Mitchell

Preface

There is a need for a comprehensive presentation of hip arthrography, dealing with all aspects but particularly with the development of the technique and evaluation of the findings. Although papers have been appearing occasionally in some medical journals over the last forty years or so, no book is yet available which is completely devoted to this subject. Sometimes this procedure is carried out by persons without adequate knowledge of the radiation hazards and the refinements of the modern radio-diagnostic equipment – factors which should be considered in the planning and setting up of such a technique.

For the last ten years we have been doing hip arthrography on a sub-regional basis – cases are referred for the examination not only from other hospitals in Sheffield, but also from the surrounding districts. Over 700 such examinations have now been carried out. Several papers on this subject appeared in the medical journals from these departments; I also had the opportunity to talk on various aspects of hip arthrography at several meetings, including clinicians, radiologists and radiographers.

I am frequently approached with requests for information about the technique as now established in Sheffield. To meet such demands, I decided to produce this monograph, which aims at providing technical details and practical advice for establishing such a radiological investigation and giving guidelines for interpreting the findings. Although many illustrations are included, this book is not intended to be a comprehensive atlas, but many different conditions as possible have been illustrated. It is hoped that it would be of particular interest to the trainee radiologists and orthopaedic surgeons, and to the senior radiographer who may be required to participate in such an investigation.

Throughout this book, special emphasis is given to radiology, but those chapters dealing with the interpretation of the radiological findings have a strong clinical bias, underlined by illustrative cases, to stress the relevance of such a radiological examination to certain clinical problems.

Acknowledgements

It would be impossible to try to mention by name all those who directly or indirectly helped me to produce this book. I am grateful to all orthopaedic surgeons and paediatricians who referred their patients for hip arthrography and for allowing me to refer to the case notes. The correlations between clinical, radiological and operative data were facilitated by their close co-operation. I would specially like to acknowledge my gratitude to Mr Evan Price, F.R.C.S. for his support and encouragement over the years.

No radiological technique can be successful without the co-operation of the radiographers. I would like to take this opportunity to thank the radiographic staff at both Northern General Hospital and Lodge Moor Hospital for their skilful and willing assistance. I would like to mention particularly Mr E. Higginbottom, H.D.S.R., Superintendent Radiographer, for his help with the section on the radiographic procedure. I am also grateful to Mr J. B. Williamson, Medical Artist, Northern General Hospital, for the line drawings accompanying the radiographs which form a major contribution to the book. Also thanks are due to Mr Alan Robinson, Senior Physicist, Sheffield Area Medical Physics Department, for his advice and help with the radiation measurements and to Mr K. Trout, Regional Statistics and Medical Records Officer, Trent Regional Health Authority, for extracting the figures from the *Report on Hospital In-patient Enquiry.* The prints of the radiographs are produced by Mr A. K. Tunstill, Chief Medical Photographer, Sheffield Area Health Authority (Teaching), to whom I am deeply indebted.

It is difficult to give full credit to all the papers which have been read in preparation of this booklet; the references given represent only a small indication of my debt to the work of others; I am especially grateful to Mr George P. Mitchell, Orthopaedic Surgeon, Edinburgh, for his helpful and constructive criticism of the manuscript and for allowing me to use Figure 7 and finally for agreeing to write the Foreword. I am also indebted to Dr M. B. Ozonoff, M.D., Director of Radiology, Newington Children's Hospital, Connecticut and to Dr Eduardo A. Salvati, M.D., Hospital for Special Surgery affiliated with the New York Hospital, for their helpful suggestions on the sections on arthrography in Perthes Disease and following Hip Replacement, respectively. Figure 45 appears by courtesy of Dr Ozonoff while Dr Salvati very kindly contributed Figure 46. I am also grateful to Dr Dennis Stoker of The Institute of Orthopaedics, London, by whose courtesy Figure 47 is included.

Some of the material and several of the illustrations have previously appeared in the following publications: *Radiography, Clinical Radiology, Archives of Diseases in Childhood, X-Ray Focus.*

I wish to express my thanks to the editors and publishers of these periodicals who have so generously given permission for this material to be included in this book.

Finally I would like to pay a special thankful tribute to my secretary, Miss Susan Tingle, for her patience and assistance with the preparation of the manuscript.

1 Introduction

1.1 Development of hip arthrography

Hip arthrography is not a new investigation; scattered reports on various aspects of the examination have appeared in the literature, especially from Europe, over the last forty years. One can hardly consider this subject without recognizing the outstanding work of Severin (1939, 1941, 1950), particularly with regard to the interpretation of the radiographic findings in the normal and in the application of the investigation in hip dysplasia.

In spite of the wide information that can be gained from such a simple and safe investigation, its use has fluctuated considerably; some people (Kenin and Levine, 1952) believe that it has not been more widely adopted because of the technical difficulties encountered. Before the introduction of image-intensification fluoroscopy, arthrography was carried out blindly, usually by the orthopaedic surgeon in the operating theatre; when some contrast was injected, few radiographs of the hip area were taken. The information gained from such a technique was limited, besides there used to be a high rate of failure in puncturing the joint capsule and the injection of contrast was often extra-capsular. Unpublished data compiled by Heublein and associates and quoted by Ozonoff (1973) indicate that 22% of all attempts at arthrography in Legg-Perthes disease were unsuccessful, and the rate of unsatisfactory studies in congenital hip dislocation was 17%.

In the average teaching hospital you will find that there are several Consultant Radiologists on the staff; this permits sub-specialization in radiology; the same happens in the specialized hospital, whether it is an Orthopaedic or Neurological hospital etc; the radiologist has the opportunity to concentrate and specialize in the particular field. This has helped the development of diagnostic radiology and as a result it has become highly specialized. I believe that as a principle, all radiological investigations should be performed by a radiologist, and this applies also to Orthopaedic Radiology. Obviously if a radiologist is going to do one of these examinations only occasionally, he is not going to do a good job technically. He will be worse off in the interpretation of the findings: after all, injecting some contrast into a joint is one thing; interpreting an arthrogram is another. I feel that such investigations should be carried out in specialist centres where such examinations are carried out regularly by interested and experienced radiologists. There is a lot of truth in the remark that 'the diagnostic value of arthrography of any joint in radiological practice is directly proportional to the interest and experience of the radiologist using it' — (Editorial comment, Year Book of Radiology, 1970). The specialist radiologist has now got the equipment, the facility and expertise to improve on the quality and value of arthrography. The image-intensified fluoroscopically controlled technique is now the method of choice, and it is used in most of the specialist Orthopaedic centres, and the success rate using this technique is close to 100%. Besides, the radiologist is more likely to be conscious of the radiation hazard and to take steps to cut down the radiation dose to a safe minimum.

There has also been improvement in the contrast media used. In the forties and fifties the agent most frequently used was Diodone 35% (Perabrodil 35%, Bayer). Toxic effects were more common then; within the last decade, this contrast medium has been replaced by sodium diatrizoate and meglumine iothalamate compounds which are better tolerated and give denser shadows at comparable concentration (Table 1).

Table 1   Contrast media currently in common use

Proprietary name	Manufacturer	Approved name	Weight/ volume	Iodine content $mg\ ml^{-1}$	Reference
Conray 280	May & Baker	Meglumine iothalamate	60%	280	Grech, 1972
Hypaque 25% 45%	Winthrop	Sodium diatrizoate	25% 45%	150 270	Astley, 1967 Katz, 1968
* Renografin 60	Squibb and Sons	Meglumine diatrizoate Sodium diatrizoate	52% 8%	290	Ozonoff, 1973
Dimer X	May & Baker	Meglumine iocarmate	60%	280	Chapter 1

* Renografin 60 used to provide 60% meglumine diatrizoate but it has since been reformulated as above which is the same as Urografin 60% (Schering Chemicals Ltd.).

1.2   Current contrast media

The contrast media now used in arthrography are usually intravascular agents which are iodinated organic compounds; these are water-soluble and of comparatively low viscosity in solution. On injection in the synovial cavity, they render the joint space radio-opaque by virtue of the radio-density of their iodine content. This degree of opacification is directly proportional to the iodine content and to the amount of contrast medium injected. The agents that are used nowadays for this examination are either sodium or meglumine salts or a mixture of both (Table 1).

Such contrast material is readily absorbed from the joint itself, and also from the soft tissue about the joint where it results from extravasation following manipulation, and it is excreted by the kidneys. Absorption is so rapid that the outline of the arthrogram becomes blurred in about ten minutes; within half an hour from injection the contrast in the joint is appreciably decreased and only a faint hazy halo around the femoral head remains; it disappears completely in about one hour after the injection (Fig. 1, a and b).

Absorption of the contrast medium is taken as equal to the rate of disappearance from within the joint. Such a rapidity of 'absorption' allows you a limited time for screening and radiography; it might not allow time to repeat a series of radiographs if the initial ones are spoilt for any reason. Consequently, it was considered worth while to try meglumine iocarmate (Dimer X) which was originally introduced for lumbosacral radiculography but has more recently been described in knee arthrography (Roebuck, 1976). Roebuck (1976) considers Dimer X vastly superior to the ordinary monomeric compounds; he is particularly impressed with the slowing down in the rate of absorption into the surfaces of the cartilage thus reducing the fuzzy appearance which is the bugbear of monomeric compounds. Recently, we have used Dimer X in ten cases of hip arthrography, injecting the same amount as with the other agents. We have found that the rate of disappearance from the joint is roughly the same as for the other media mentioned above. (Fig. 1, c and d). Since Dimer X is appreciably more expensive than the other agents (almost fivefold) and does not seem to give any obvious advantages, we have reverted to using Conray 280.

We did not see the need for adding adrenaline as recommended by Hall (1974) in knee arthrography to improve the sharpness of the contrast-outlined intra-articular structures as well as the persistence of this enhanced detail in the delayed radiographs, especially as many of the patients referred for hip arthrograms were children under one year of age.

No local or general reactions have been observed, confirming previous observations showing that this examination is relatively harmless (Leveuf and Bertrand, 1937; Heublein et al., 1952, and Mitchell, 1963). Although the contrast media in use nowadays are less toxic than the previous ones, they should not be considered as being entirely free of toxicity (Grainger, 1970). Toxic reactions are probably so rare in this examination because of:

(a)   The type of examination   – the contrast is injected in a synovial cavity and not intravascularly.

(b)   The type of patient   – usually the patient is otherwise healthy, and this may be significant since the incidence of reaction is greater in those with renal or hepatic disease, thyrotoxicosis or seriously ill patients.

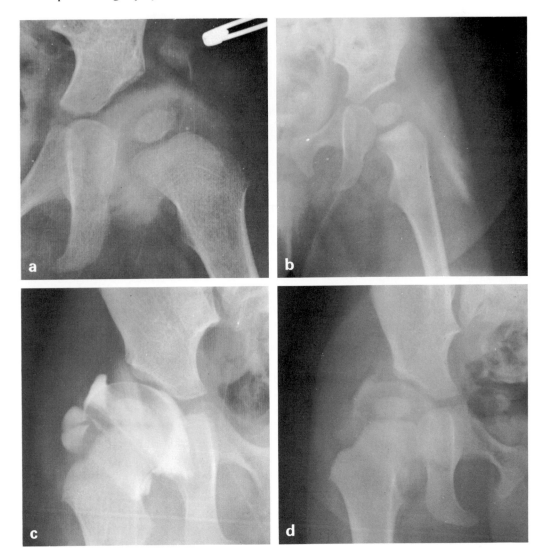

Figure 1   Absorption of the contrast medium from the joint following arthrography. Radiograph (a) was taken forty-five minutes after injecting Conray 280; it shows only a 'halo' of the contrast. It has completely disappeared one hour after injection (b). Dimer X was used in (c); again this contrast medium is almost completely gone thirty minutes after the injection (d).

(c)   The amount of contrast   – a few ml of contrast medium are usually injected and therefore the total amount of iodine is very small.

In spite of this apparent safety of the examination, every precaution must be taken to prevent adverse reaction. All patients should be questioned to exclude any history of allergy, asthma or sensitivity to iodine. If there is such a history the clinical indications for the investigation should be reviewed with the clinician responsible for the patient, and the value of the information to be gained weighed against the potential risk. If the examination is still considered necessary, it should be carried out under hydrocortisone cover and one should ensure that full resuscitation facilities are available in the X-ray room where the investigation is being carried out.

From the reports available, it appears that it makes little difference which preparation is chosen.

1.3   Basic principles

There are certain criteria which must be followed in carrying out hip arthrography.

1.3.1   *Basic principles before the examination is undertaken*

(1) The indications for the examination should be well-founded and the problems to be solved well-defined.
(2) There should be close co-operation between the radiologist and the clinician.
(3) Not all examinations should be performed in a standardized way; some will have to be varied according to the information required. Every examination should be performed in such a way as to yield a solution to the problem.
(4) If possible the procedure should be undertaken as a combined exercise by the radiologist and the clinician; if this is not practicable, then the examination should be adequately documented so that the case can be thoroughly discussed afterwards.
(5) The investigation should be carried out in the X-ray department and the equipment should be adequate and in good condition.
(6) The examiner should be competent and all possible steps should be taken to keep down the radiation dose to a minimum.

1.3.2   *Basic principles for the procedure of arthrocentesis*

(1) The procedure must be undertaken under strict aseptic conditions.
(2) Anatomical landmarks should be utilized for orientation and major vessels and nerves avoided.
(3) Unnecessary trauma to articular cartilage should be avoided.
(4) The joint should be in a position to give maximum relaxation of the capsule to facilitate puncturing of the joint capsule.

6 Hip arthrography

References

Astley, R. (1967), Arthrography in congenital dislocation of the hip. *Clinical Radiology*, **18**, 253–260.

Grainger, R. G. (1970), Contrast media, In: *Modern Trends in Diagnostic Radiology.* (J. W. McLaren, Ed.), Series IV, Butterworth, London.

Grech, P. (1972), Arthrography in hip dysplasia in infants. *Radiology*, **38**, 172–179.

Hall, F. M. (1974), Epinephrine – enhanced knee arthrography. *Radiology*, **111**, 215–217.

Heublein, G. W., Greeme, G. S. and Comforti, V. P. (1952), Hip joint arthrography. *American Journal of Roentgenology*, **68**, 736–748.

Holt, J. F., Whitehouse, W. M. and Latourelle, H. B. (Eds.) (1970), *Year Book of Radiology.* Year Book Medical Publishers Inc., Chicago.

Katz, J. F. (1968), Arthrography in Legg–Calvé–Perthes Disease. *Journal of Bone and Joint Surgery*, **50–A**, 467–472.

Kenin, A. and Levine, J. (1952), A technique for arthrography of the hip. *American Journal of Roentgenology*, **68**, 107–111.

Leveuf, J. and Bertrand, P. (1937), L'arthrographie dans la luxation congenitale de la hauche. *Presse med.*, **45**, 437–444.

Mitchell, G. P. (1963), Arthrography in congenital displacement of the hip. *Journal of Bone and Joint Surgery*, **45-B**, 88–95.

Ozonoff, M. B. (1973), Controlled arthrography of the hip: A technic of fluoroscopic monitoring and recording. *Clinical Orthopaedics*, **93**, 260–264.

Roebuck, E. J. (1976), Contrast media in knee arthrography. *British Journal of Radiology*, **49**, 287.

Roebuck, E. J. (1976), Personal communication.

Severin, E. (1939), Arthrography in congenital dislocation of the hip. *Journal of Bone and Joint Surgery*, **21**, 304–313.

Severin, E. (1941), Arthrograms of hip joints of children. *Surgery, Gynaecology and Obstetrics*, **72**, 601–604.

Severin, E. (1950), Congenital dislocation of the hip – development of the joint after closed reduction. *Journal of Bone and Joint Surgery*, **32–A**, 507–518.

2 Radiological anatomy

The radiological diagnosis of disorders of the hip joint is based on the understanding of the normal development, anatomy and function of the joint and the measurements of certain relevant anatomical structures of the joint. Only anatomical aspects that are relevant to arthrography are discussed in this section.

The hip is a synovial joint which presents the following characteristic features:

(a) It is formed by two separate bony components – the acetabulum and the femoral head – which are only attached together by the ligamentum teres.
(b) The opposing bony surfaces are lined by articular cartilage.
(c) There is a joint cavity containing synovial fluid.
(d) The joint is enclosed in a fibrous capsule.
(e) The inside of the joint, except for the articular cartilage, is lined by the synovial membrane.

2.1 Osteology

The hip is the most stable ball and socket joint in the body. It consists of the femoral head, which forms roughly two-thirds of a sphere, and the acetabulum, a cup-shaped socket made up of the iliac, ischial and pubic parts. Both articulating components are covered with articular cartilage; in the acetabulum it is found only on the horseshoe-shaped part as the depressed, roughened fossa is actually non-articular and consequently not covered with articular cartilage. This fossa contains a fibro-elastic fat pad and the flat ligamentum teres which is attached to the fovea capitis on the femoral head. The depression of the fossa together with the elasticity of the fat pad allows free movements to the ligamentum teres without it being compressed between the articulating surfaces.

Estimation of the depth of the socket is sometimes important. The ratio of the depth of the acetabular socket to its length is called the *acetabular index*. This is calculated from the measurements obtained from an antero-posterior view of the hip joint as follows:

$$\text{Acetabular index} = \frac{\text{depth of socket}}{\text{length of socket}}$$

In Fig. 2, a, this measures 32/57 mm or 0·56. If this index is less than 0·5, as in Fig. 2, b, it would indicate that the acetabulum is shallow.

The head of the femur is partially contained within the acetabulum. At birth (Fig. 3, a) the femoral head is completely cartilaginous; ossification starts about the age of six months as a single ossific centre, usually situated in the centre of the spherical head. This ossification advances gradually (Figs. 3, b–e) and by the age of thirteen years the whole of the head is ossified, except for the articular layer (Fig. 3, f).

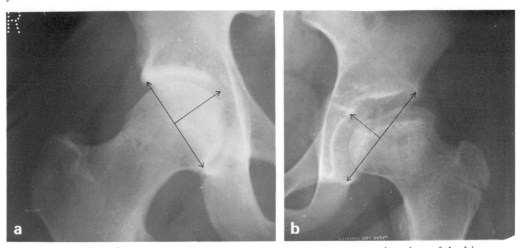

Figure 2   Estimation of the acetabular index from an anteroposterior view of the hip joint. This is obtained by the ratio of the depth to the length of the acetabulum. In a normal patient (a), this is estimated at 0·56; while in (b), a case of healed Perthes disease, it is found to be 0·34.

However, the three centres of the upper end of the femur – the capital, the greater and lesser trochanteric centres – do not fuse to the rest of the bone until about the eighteenth year (Fig. 3, g).

2.2   The labrum

The transverse ligament, which runs across the acetabular notch, continues as a ring of tough but pliable fibro-cartilage and it is attached to the brim of the acetabulum. This deepens the socket and makes the joint more stable. The labrum is particularly prominent on the postero-superior aspect of the acetabulum where it is lined by synovial membrane on both its superficial and deep surfaces. As a result this part makes a free margin, called the *limbus* (Fig. 4); it is mobile and as a result it can be inverted into the joint cavity in congenital dislocation of the hip.

2.3   The capsule

This is a strong closely fitting fibrous sleeve attached proximally around the brim of the acetabulum, the labrum and the transverse ligament. It covers the head and neck

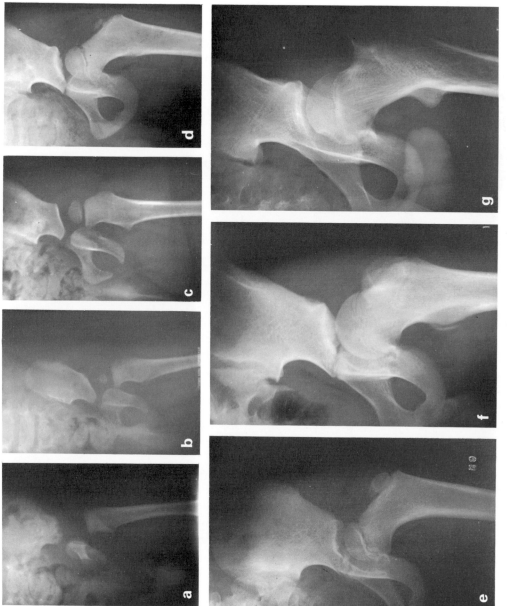

Figure 3 Series of antero-posterior radiographs of the hip joint, showing development and maturation with age. (a) in a child of 3 months; (b) 8 months; (c) 1 year; (d) 5 years; (e) 9 years; (f) 13 years; (g) 18 years.

of the femur, and reaches to the intertrochanteric line anteriorly and posteriorly to the junction of the middle and distal thirds of the neck of the femur. The capsule is constricted around the narrowest part of the neck by the *zona orbicularis* – a bundle of deeply placed circular fibres which strengthens the retaining power of the labrum producing an hour-glass constriction within the capsule (Fig. 4).

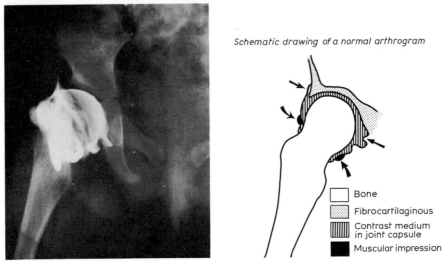

Schematic drawing of a normal arthrogram

Bone

Fibrocartilaginous

Contrast medium in joint capsule

Muscular impression

Figure 4   Normal hip arthrogram in a seven-month-old child with a schematic drawing to show the fibro-cartilaginous structures. The short straight arrow points to the 'rose-thorn' projection; the long arrow to the 'groove' caused by the impression on the synovial membrane by the transverse ligament, and the curved arrows point to the bundle of circular muscular fibres which cause a constriction of the capsule – the 'zona orbicularis'.

2.4   The synovial membrane

The synovial membrane lines all the surfaces of the interior of the joint, except where there is either articular cartilage or fibrocartilage. The synovial membrane arises not from the edge of the limbus but a bit higher up, leaving a triangular-shaped pocket between itself and the outer side of the limbus (Fig. 4); when this is filled with contrast medium it is usually known as the '*rose-thorn*' projection. The synovial lining extends to the neck of the femur and covers its anterior aspect completely. It is loosely attached and consequently it is distensible. As the psoas muscle proceeds downwards into the thigh, it passes behind the inguinal ligament and in front of the capsule of the hip joint, from which it is separated by a bursa. Occasionally this psoas bursa communicates with the cavity of the hip joint. The relationship of the psoas muscle to the hip joint is well shown in Fig. 5; the fusiform shadow (see arrows) represents the psoas which is outlined by the direct injection of the contrast inside its sheath.

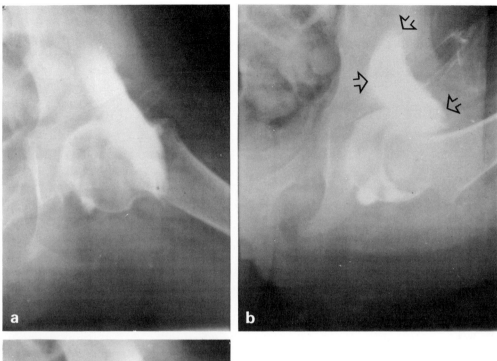

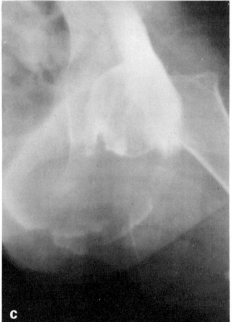

Figure 5   Three different views of an arthrogram on a girl of ten months showing a normal left hip, with injection of contrast into the psoas sheath to show the relationship of this muscle to the hip joint. The arrows point to a fusiform shadow representing the psoas sheath, which moves with the joint. In (b), 'frog' view, it is shown to lie anteriorly to the joint capsule.

Inside the synovial membrane there is a film of synovial fluid. This acts as a joint lubricant between the articulating surfaces; it also provides nutrition for the avascular articular cartilage, and most probably takes part in the dissipation of the heat generated from its movements and action and other end-products of metabolism.

2.5   Width of the hip joint space

In the adult, slight joint narrowing could be an early sign of articular cartilage erosion or degeneration, but this cannot usually be seen so early on an antero-posterior radiograph, because in such a projection the hip joint socket cannot be displayed in its full extent because of the slope of the joint. It is argued that this can be improved upon by using Lequesné technique, which gives a better view of the antero-superior aspect of the roof and the postero-medial part of the hip joint space. We have found this technique to be difficult to perform in some patients, and it also entails a high object-film distance which would give appreciable magnification in spite of using a large target-film distance of 150 cm. This also leads to an increased exposure on account of distance and the focus may need changing from fine to broad.

We have found the lateral view for the head of the femur satisfactory (Fig. 6). The patient is turned on to the side to be examined with flexion at both hip and knee joint, the pelvis being tilted for 45° backward and the good limb raised and supported on pads. From a statistical analysis involving the Lequesné technique, Zwicker *et al.* (1969) found the mean joint space of twenty-six normal patients varied from 3 mm to 4 mm. Using a similar method Lusted and Keats (1972) obtained the same results for the normal adult. Approximately the same results were obtained using a simple lateral view as described above. All these measurements were obtained on plain radiographs; the joint space in these studies not only included the synovial space but also the articular cartilages. It is argued that when articular degeneration is present, joint narrowing can be noted early with either of these views.

The width of the joint space, as seen on the arthrogram, does not correspond completely with that seen on the plain radiograph. The arthrogram gives the true joint space, i.e. that potential synovial space which gets filled with contrast medium; while, on the plain radiograph, the space is considered to be the gap between the ossified bony parts. In the adult, such a gap includes the articular cartilages while in the young child most of the femoral head could still be cartilaginous.

2.6   Ontogenesis

Primary ossification centres appear in the long bones from the eighth week onwards of intra-uterine life. As a result the shaft of the femur is already ossified at term. However, the head, trochanter and neck are still a cartilaginous mass and consequently not radio-opaque (Fig. 3, a). This cartilaginous block is finally ossified.

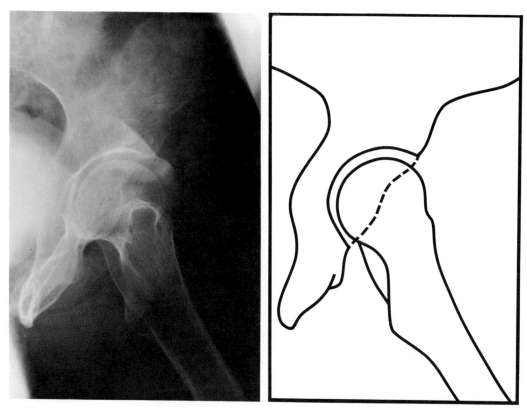

Figure 6   Estimation of the hip joint space. This was found easier to calculate from a lateral view for the head of the femur. In a normal adult, this was found to vary from 3 to 4 mm. This gap includes the actual joint space and the articular cartilage.

from three secondary centres. The times of appearance of these three centres are somewhat variable, but they should be expected to be present as follows:

(a) Centre for the head – by the age of six months (Fig. 3, b);
(b) Centre for the greater trochanter – by the age of four years (Fig. 3, d);
(c) Centre for the lesser trochanter – by the age of 13 years (Fig. 3, f).

All these three centres are fused to the femoral shaft by the eighteenth year of life (Fig. 3, g).

At birth, the acetabulum is a cartilaginous cup. Ossification of the acetabular cartilage begins as two separate centres which become fused as ossification spreads. This centre forms a considerable part of the articular surface of the adult bone along the junction of the tri-radiate stem of the cup of cartilage and along the margin of the cup.

Laurenson (1965) studied the development of the hip joint by prenatal experimental arthrography on fetuses ranging in age from fourteen weeks to term. He showed that all the articular structures identifiable on the arthrogram of the normal hip of the young infant are present in the hip of the fourteen-week-old fetus. No extraordinary change in the depth or shape of the acetabulum was noted. The acetabular roof was found to differ throughout fetal development – at fourteen weeks the acetabular roof was entirely cartilaginous; by the twenty-fifth week the apex of the lower border of the centre of ossification had moved medially, while bone had spread laterally and inferiorly into the base of the limbus. At birth, the apex of the centre of ossification was two millimetres below the level of the limbus. This definitive shape of the bony roof of the acetabulum had become more obvious and there was a corresponding reduction in the extent of the cartilaginous roof. Throughout this development, the limbus, the zona orbicularis and the ligamentum teres remain prominent features of the arthrogram.

Normally, during growth, there is change in the axial relationship of the hip joint. The angle formed by the axis of the neck of the femur with the axis of the shaft changes from 160° to about 125°; at the same time the torsion angle changes from about 35° anteversion to about 15° anteversion.

2.7   Movements of the hip

The construction of the hip joint is such as to permit a wide range of movements which are described as polyaxial, i.e. the femur is capable of motion around an indefinite number of axes with an imaginary common centre. These movements include flexion and extension, adduction and abduction, circumduction and rotation. Although the femoral head is not truly spherical, nor is the acetabulum perfectly concave, these rotatory movements can be considered as being *concentric* – they occur around this common centre. For this reason there is very little difference in the width of the joint space in the various positions of the normal hip; the articular surfaces between the acetabulum and the femoral head show and maintain '*parallelism*' between them throughout such movements.

References

Laurenson, R. D. (1965), Development of the acetabular roof in the fetal hip. An arthrographic and histological study. *Journal of Bone and Joint Surgery*, **47–A**, 975–983.

Lequesné, M. (1969), Die Erkrankungen des Hüftgelenkes bein Erwachsenen. *Folia rheumatologica*, **17a**, Documenta Geigy.

Lusted, L. B. and Keats, T. E. (1972) *Atlas of Roentgenographic measurements*, 3rd edn. Year Book Medical Publishers Inc., Chicago.

Zwicker, H., Münzenberg, K. J. and Düx, A. (1969), Huftgelenkveränderungen bei Morbero Cushing. *Fortschr. Geb. Röntgenstrahlem*, **111**, 693–697.

3 Technique

Briefly, the examination consists in injecting contrast medium into the joint space and then manipulating the joint under television screening, and finally recording the findings. There are four stages in the procedure which will be considered in some detail.

3.1 Preparation of the patient

Before arriving in the X-ray department
(a) The patient is admitted in the hospital for the day.
(b) In the beginning we often found that the patient or parents did not realize what the examination entailed. Consequently an explanatory note (Appendix I) was prepared and sent to the patient together with the appointment for the examination as a day admission. We found such prior information about the examination helpful and avoided unnecessary misunderstanding.
(c) A consent form, same as that used for any other operative procedure, is duly completed.

In the X-ray department
(d) If the patient is under 10 years of age, the examination is carried out under general anaesthesia; with a co-operative, older patient local anaesthesia is adequate – the site of the insertion of the needle is infiltrated with a two per cent solution of any of the conventional parenteral local anaesthetic agents.
(e) If the patient is a small child, once anaesthesia is induced, the infant is put on the pelvis holder (see Section 5.2).
(f) The patient, lying supine on the radiographic table, is put with the hip in the position of rest – which is approximately 10° flexion, 10° abduction and 10° external rotation (Tronzo, 1973). This position allows total capsular slackness and maximal joint capacity with complete muscular relaxation.
(g) If the patient is a child in a plaster spica, anaesthesia is first induced; the plaster cast is then removed and the child is put on the pelvis holder. After the examination the child is put back in the bivalved plaster spica.

3.2   Arthrocentesis

The skin area is cleansed and sterilized with an effective bactericidal agent and the examination is carried out under full aseptic technique, using a pre-assembled arthrography tray (Appendix 2).

A recent radiograph of the hip joint is first studied; if none are available a control radiograph is taken. The osseous landmarks and femoral artery are localized by palpation and the approach to the joint cavity is decided upon.

There are at least four possible routes to enter the hip joint:

(a) Superior approach – the joint is entered from above and some workers, including Mitchell (1963), prefer this approach, particularly if there is an excess of fluid. They claim that any leakage that may occur from the joint flows upwards and does not spoil the arthrographic findings (Fig. 7).

(b) Inferior approach – the needle is inserted underneath the adductor tendons and is advanced cephalad to hit the femoral head or the neck close to the head (Fig. 8, a).

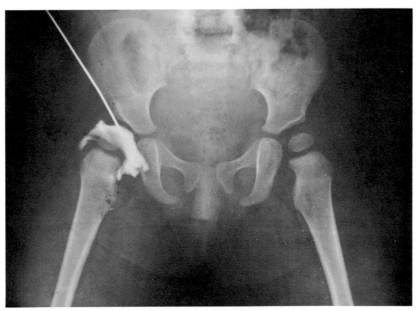

Figure 7   Possible route to enter the joint space. Superior approach – it is maintained that it avoids obscuring of the joint by escaping fluid. This right hip arthrogram shows an inverted limbus very well.

(c) Anterior approach – the needle is advanced antero-posteriorly usually aiming to enter the joint at the lateral aspect of the femoral head (Fig. 8, b).

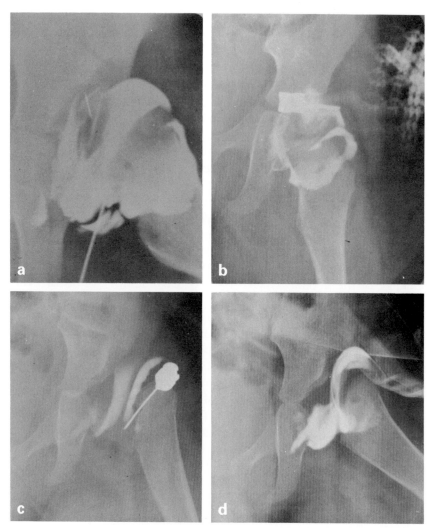

Figure 8 Other possible routes to enter the joint space. (a) Inferior approach – lumbar puncture needle is inserted underneath the adductor tendons and is pushed upwards to hit the femoral head close to the neck. (b) Anterior approach – aiming the needle to enter the joint into the synovial recess over the neck. This case was investigated when we first introduced this technique and we were using ordinary stainless steel serum needles. (c) Anterior approach (medial). When there is marked lateral displacement of the femoral head, as in this case, it is easier to enter the joint space medially to the femoral head. Monoject 220 disposable needles are now used; note how much finer the needle is. (d) When the joint space is adequately filled with contrast, it is shown that this displacement is due to hour-glass constriction of the capsule.

(d) Lateral approach – the needle is introduced laterally, directly above the tip of the greater trochanter and moved along the superior surface of the neck until the joint is entered.

Most workers, including Severin (1939), Heublein, Greeme and Comforti (1952), Kenin and Levine (1952), Crenshaw (1963) and Ozonoff (1973) seem to agree that the anterior approach is the best one. Although one always gets a bit of leakage of contrast from the puncture hole when the needle is removed and the joint manipulated, this does not obscure the area nor interfere with the findings (Fig. 11).

It is advisable to use short-bevelled needles, otherwise the needle might be only partially inside the joint space. Monoject 220, disposable spinal needles with stylet from Sherwood Medical Industries (Florida), are found to be very suitable. Gauge 22, $1\frac{1}{2}$ inch long are used on infants and young children, while gauge 20, $3\frac{1}{2}$ inch ones are preferred for older children and adults. Obviously, if other routes for entering the joint are adopted, like the inferior or the lateral approach, one would have to use the longer needles even on infants. The stylet is probably of some advantage in the prevention of blockage of the needle by soft tissue or cartilaginous plugs or even blood clots.

Image-amplified fluoroscopically controlled arthrocentesis is now considered the method of choice. The basic advantage of fluoroscopic guidance is the ability of the operator to place the needle precisely within the capsule in the site of preference. As has been previously stressed by Ozonoff (1973), one should try to avoid needling the growth plate on the epiphysis or damaging the labrum. The method we follow is that after palpation of the hip joint and localization of the major vessels, a needle is inserted in the skin and its position, in relation to the femoral head, is checked by fluoroscopy. If needed it is adjusted until it is aimed to hit the femoral head tangentially at its supero-lateral aspect. Once it is considered that the needle is pointing in the right direction, it is advanced until it enters the capsule. Ideally, one should try to place the needle into the synovial recess over the neck. Once it is ascertained that the needle is properly positioned, the injection of contrast medium is made. Fluoroscopic guidance in puncturing the joint space is important because it simplifies and quickens the procedure.

It is often recommended that one should insert the needle one finger breadth or one centimetre lateral to the femoral artery along Poupart's ligament. This applies only to young children, say under one year; as the child grows the site for needle insertion is found further away from the femoral artery. This relationship of the major vessels to the hip joint is shown in Figure 9.

Also, when the femoral head is grossly dislocated, it might be easier to enter the joint space medially to the femoral head (Fig. 8, c and d) – especially if such a dislocation is associated with an hour-glass deformity of the joint lining, when the capsule is pulled medially and it is so stretched across the lateral aspect of the femoral head that the potential space in this region is limited further.

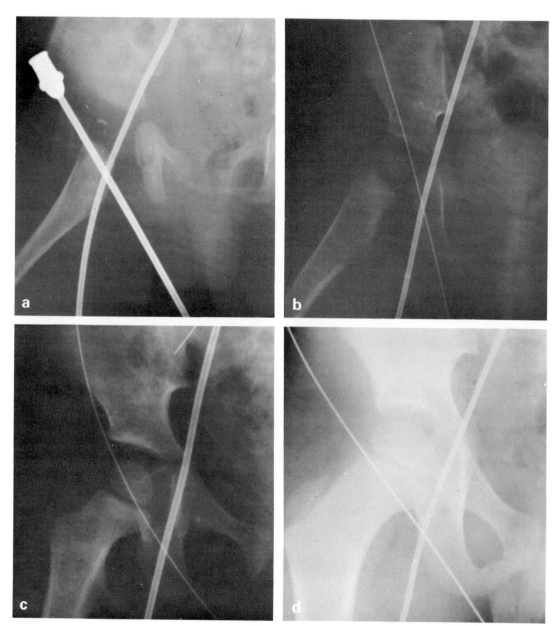

Figure 9　Relationship of the femoral head to the femoral vein. Radiographs of four patients, (a) aged 7 weeks, (b) 10 months, (c) 3 years and (d) 8 years, with catheter in femoral vein and marker along the inguinal ligament. It is obvious that the rule that the needle should be inserted one cm or one finger breadth lateral to the femoral artery, applies only to infants.

3.3 Injection of contrast

When it is felt that the joint capsule has been entered, a check is made by injecting a few millilitres of normal saline. If the joint has been successfully entered, on injecting saline, this will be under pressure and on releasing the plunger of the syringe the saline will flow back into the syringe. One can get an idea about the state of the joint from the ease and amount of saline one can inject before this flow back becomes apparent; a lax and baggy capsule will take much more saline than a normal hip.

If free synovial fluid is present, this will run up along the needle on puncturing the joint; if such synovial fluid is present in excessive amount, it should be aspirated before the contrast agent is introduced.

Once it is ascertained that the needle is properly positioned inside the joint, the saline syringe is replaced by another containing the contrast medium, and few millilitres of the contrast are injected. In young patients, 1 to 2 ml is now considered adequate; if there is a lax capsule, or frank dislocation and in all adults, the capacity is increased, and in such cases one needs 3–4 ml or more of contrast. The amount of contrast required is decided by the fluoroscopic appearances. Too much contrast might obscure rather than clarify the findings; too much injected fluid blurs the outline. Besides, the more fluid injected the more the capsule is distended and the more marked is the extra-capsular leakage during manipulation. Some workers dilute the contrast medium to avoid too high a density; if, however, normal saline is used to ascertain proper positioning of the needle, this will help to dilute the injected contrast.

When the contrast medium is injected, a quick check is made by television screening to make sure that this is adequate; some workers inject the contrast medium under fluorscopy. When the injection is completed, the needle is removed and some flexible collodion is applied to the skin puncture.

One has to make sure that the needle does not come out of the capsule while changing the syringes, the contrast syringe for the one containing saline. One should check after a trace of contrast is injected by fluoroscopy; if it is in the joint properly it will quickly flow around the periphery in a very characteristic fashion that leaves the radiologist in no doubt that the injection is properly made; if however the needle is not correctly placed, the contrast medium will remain at the site of injection without spreading and, from this appearance, it is obviously extra-capsular and lying in the soft tissue (Fig. 10).

3.4 Radiological examination

Once the needle is removed, the radiological examination is carried out and the findings recorded. The method that is followed is dictated by the condition; it will have to be varied according to the information that is required. The examination makes it possible to study not only the anatomy of the joint but also the dynamics

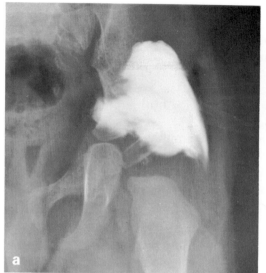 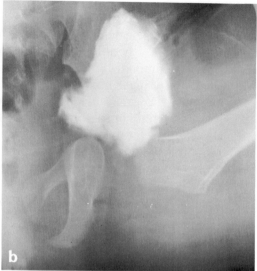

Figure 10 Extra-capsular injection. Attempted left arthrogram on a six-month-old child. The contrast medium did not spread in the characteristic fashion and remained at the site of injection shown in neutral position (a) and also in frog position (b).

across it. This shows important structures that cannot be seen on the plain radiograph, because they are not radio-opaque, especially in early childhood when a large part of the hip joint is cartilaginous and therefore not radiographically demonstrable. By outlining these structures with the contrast medium, they are rendered visible. Secondly, the joint can then be manipulated and observed under television screening and the dynamics of the joint assessed throughout the full manipulative range of movements. The relationship of the bones forming the joint in various positions, the range of mobility, the competence of the supporting structures and the stability of the joint can be observed. Consequently the radiographs taken depend on the condition that is being investigated; obviously more thorough recording is needed if not only the anatomy, but also the functional state is being assessed.

We take all spot films with the undercouch tube. Some radiologists including Astley (1967) and Ozonoff (1973) prefer overcouch exposures; with the equipment available the radiographic quality of our undercouch exposures is satisfactory; also we document most of the procedures on videotape and prefer to record while we are actually screening (Grech, 1972). Overcouch radiography, whether from a ceiling mounted tube or a floor mounted one, would entail some delay in bringing the overcouch tube across and removing the explorator.

Teamwork between the radiographic and radiological staff is essential. Though there is no need for undue haste, the radiography should be carried out with deliberate speed, because within fifteen minutes or so the outline of the arthrographic shadows becomes hazy. The spot films once exposed should be processed and inspected straightaway, in case any need repeating or an apparent lesion not previously noted needs further views.

Such a developed technique has now been applied in our departments for the last ten years, and we have not had a failed attempt over the last five hundred consecutive arthrograms. The method has been found suitable for both children and adults. Sometimes bilateral hip arthrography is needed (Fig. 11) – we usually carry these out at one sitting; preferably, injections should be made simultaneously on both sides in order to cut down on radiation; if there is some difficulty in entering the second joint, then the examination should be continued with the first and another attempt made to enter the second joint afterwards; but absorption is so rapid that sharp and detailed radiographs are often not obtainable fifteen minutes after the injection.

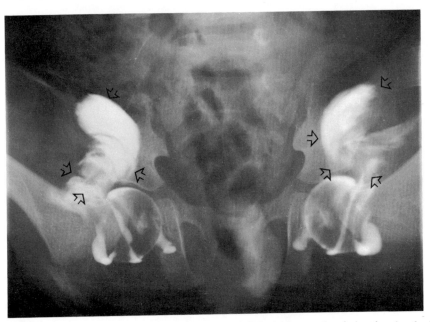

Figure 11   Bilateral hip arthrogram on a child of 18 months, injecting both sides simultaneously, showing both hip joints to be normal. There is a fair amount of leakage of contrast medium due to the manipulation (see arrows), but this does not obscure the findings.

3.5   Radiographic procedures

Originally, before routine video-tape recordings were undertaken, spot films were taken as and when required, during manipulation under screening control.

When the video-tape recorder became available this arrangement was continued.

However the video-tape playback was broken up by pick-up images between each separate recording. This was due to the video-tape requiring time to take up tension and the recording head needing time to achieve the correct recording speed. It was decided to keep both imaging sequences separate; first the dynamic video recording followed by spot films. To reduce screening time further, the image intensifier is centred to the hip joint or joints before the X-ray tube is energized during the initial stages of the examination and before the introduction of the contrast medium. If additional filtration has been added to the table top (i.e. tinplate) to reduce 'whiteout' when screening very small babies, it is essential to remove the filter before radiography is commenced. The duodenal cap localizer is used for collimation since this gives a field size of 12·5 cm × 10 cm which is adequate for children up to 10 years of age. For bilateral hip arthrograms and all others an appropriate field size is selected. On most modern screening tables there is a field size limiting device on the undercouch tube collimator which automatically maintains field size irrespective of tube-screen distance. A preliminary radiograph is necessary so as to guarantee the exposure factors, for the fairly rapid series of radiographs is usually completed by the time the first contrast film is ready for inspection.

3.5.1 *The video-tape sequence*

When there is adequate delineation of the hip joint, the video-tape recorder or disc recorder is switched on, the tube and image intensifier centred over the appropriate hip joint. Continuous screening and recording takes place during manipulation of the hip joint.

The manipulative technique is subject to considerable variation, but the general principle is to study the full anatomy of the joint. In certain conditions like hip dysplasia, the examination of the dynamics of the joint forms an integral part of such a study. Consequently, although the complete sequence of positions should be followed in such disorders where the function of dynamics of the joint may be affected, the examination can be curtailed in other conditions, as in arthrography following total hip replacement. Details are given here for the full examination as used in case of hip dysplasia.

The sequence starts with the patient lying supine on the radiographic table with legs in a neutral position. The knees are then pushed down onto the table to hyper-extend the hip joint. The leg is then abducted to an angle of forty five degrees, first internally rotated and then externally rotated. This is again repeated with approximately sixty degrees abduction.

The leg is now returned to the neutral position and it is pulled caudally (traction) and pushed cephalically (telescopy) so that the joint is alternately under extension and compression. The affected leg is then adducted and crossed over the other leg and pressure is applied to the bent knee. The leg is then returned to the neutral position. The leg is then flexed at both hip and knee joints and pressure is applied to the bent knee caudally. Finally the hip is put in a 'frog' position which is obtained by flexion of the hip and knee, followed by rotation of the limb laterally from the hip joint through approximately sixty degrees.

Where doubtful or very transient pathology is noted repeat sequences may be recorded.

3.5.2   *The spot-film technique*

Using an undercouch tube, all radiographic projections will be postero-anterior. The inherent filtration of the table-top attenuates the primary beam before irradiating the patient. Because of the short tube-object, object-film ratio, a fine focal spot film is essential, preferably 0·6 mm. It is not necessary to record all the different projections used in the dynamic recording. Only those required to establish pathology or normality are taken on spot films. For babies, exposure factors of 16 mA.s and 65 kVp have been used with high definition screens and fast films. The same exposure factors can be used for all projections except the 'Ortolani' where an increase of 5 kVp and a further 4–10 mA.s is required. In all cases a grid is used. A progressive increase in exposure factors will be required for older children and adults. Although a combination of rare-earth screens and fast films will dramatically reduce exposure levels, quantum mottle may well limit diagnostic quality, especially with babies. Consequently a combination of rare-earth screens and the new slower films, giving an overall increase in speed of 2 to $2\frac{1}{2}$ times the exposure required on the fast film/fast tungstate combination, are to be recommended.

3.5.3   *Spot-film sequence*

There is no need to radiograph all projections, especially if the procedure is recorded on video-tape; only those views necessary for permanently recording the pathology need be taken. However, the whole manoeuvre can be documented on the following basic spot-film sequence:

(1) Neutral position with hyper-extension of hips by pushing the knees on to the radiographic table (Fig. 12, a).
(2) Leg abducted to 45° and maximally internally rotated (Von Rosen position) (Fig. 12, b).
(3) Leg abducted fully and rotated (Fig. 12, c).
(4) Leg that is being examined is crossed over the other, and pressure applied on the flexed knee (Ortolani technique) (Fig. 12, d).
(5) Leg is flexed at both hip and knee joints and pressure is applied to the bent knee caudally (Fig. 12, e).
(6) Frog position (Fig. 12, f).

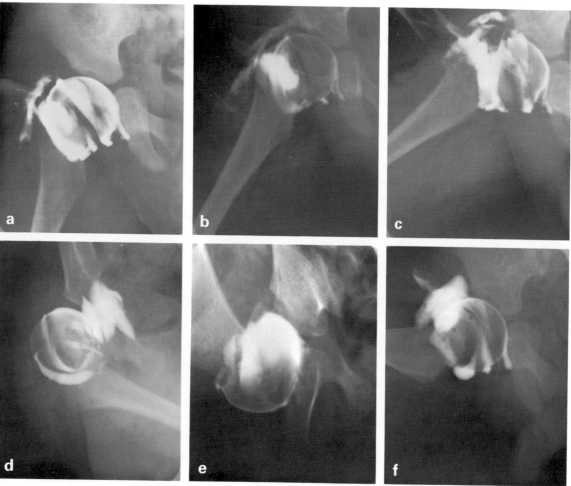

Figure 12   Spot-film sequence in a normal arthrographic study. (a) Neutral position with hyper-extension of the hip by pushing down on the knee. (b) 45° abduction with external rotation. (c) 60° abduction with internal rotation. (d) Leg crossed over to opposite side and pressure applied on bent knee. (e) Leg flexed at hip and knee and pressure applied caudally on the bent knee. (f) 'Frog' position.

Besides these basic positions, sometimes an abnormality may be shown better in some other position not included in this series. Playback of the video-tape will show the best position to demonstrate the abnormality.

Rapid film processing is of considerable advantage in producing the full series for inspection before the contrast is absorbed.

Appendix 1

Hip joint examination on a child

(1) This examination is carried out under a general anaesthetic; therefore in the interests of the safety of the child he/she must have nothing by mouth for four hours prior to the examination.

(2) Bring the child to .... Hospital, Ward ...... by 10.0 a.m. The child will be kept in hospital for the day and he/she will be able to go home about 5.0 p.m.

(3) Usually the examination takes about thirty minutes.

(4) Do not bring the child if he/she has got a cold as it is then inadvisable for the child to have a general anaesthetic. Also it is inadvisable to carry out the examination if there is a skin rash in the groin. Please ring and make another appointment.

Hip joint examination on older children/adults

(1) This examination is carried out under a local anaesthesia, but in your own interest you should not have anything by mouth for four hours prior to the examination.

(2) Please report to .... Hospital, Ward ...... by 10.0 a.m. You will be kept in hospital for the day, but you will be able to go home about 5.0 p.m.

(3) Usually the examination takes about thirty minutes.

(4) If you intend to come to hospital by your own transport, please arrange for somebody to drive you back, as it is not advisable to drive yourself back.

(5) If you cannot keep this appointment, please ring the department and arrange for another one.

Appendix 2

Trolley setting

(1) *Upper shelf – sterile*
Gowns and gloves
Dressing towels
Gauze and cotton wool swabs
5 ml syringe – for contrast
10 ml syringe – for saline
Short bevelled needles – GA $22 \times 1\frac{1}{2}$ in for young children; GA $20 \times 3\frac{1}{2}$ in for older children and adults.
One aspiration needle and quill

Two gallipots for skin antiseptic
One bowl for saline
One pair of forceps to hold swabs for cleaning the skin

(2) *Lower shelf – non-sterile*
Masks
Files
Spare needles and syringes
Saline
Contrast medium ampoules
Antiseptics
Flexible collodion

References

Astley, R. (1967), Arthrography in congenital dislocation of the hip. *Clinical Radiology*, **18**, 253–260.

Crenshaw, A. H. (1963), *Campbell's orthopaedics*, 4th edn., Vol. II, The C. V. Mosby Co., Saint Louis, U.S.A.

Grech, P. (1972), Video-arthrography in hip dysplasia. *Clinical Radiology*, **23**, 202–207.

Heublein, G. W., Greeme, G. S. and Comforti, V. P. (1952), Hip joint arthrography. *American Journal of Roentgenology*, **68**, 736–748.

Kenin, A. and Levine, J. (1952), A technique for arthrography of the hip. *American Journal of Roentgenology*, **68**, 107–111.

Mitchell, G. P. (1963), Arthrography in congenital displacement of the hip. *Journal of Bone and Joint Surgery*, **45–B**, 88–95.

Ozonoff, M. B. (1973) Controlled arthrography of the hip: A technic of fluoroscopic monitoring and recording. *Clinical Orthopaedics*, **93**, 260–264.

Severin, E. (1939), Arthrograms in congenital dislocation of the hip. *Journal of Bone and Joint Surgery*, **21**, 304–313.

Tronzo, R. G. (1973), *Surgery of the Hip Joint*, Lea Febiger, Philadelphia.

4 Difficulties, complications and contra-indications

4.1 Difficulties and complications

If the technique is carried out as detailed above, very few difficulties or complications should be encountered, and these are usually avoidable. At first, ordinary needles were used for arthrography and sometimes they got blocked. It is advisable to use a needle with a stylet and it should be introduced with the stylet in it to avoid blocking with soft tissue or blood. When the needle is reckoned to be properly situated, the stylet should be removed and sterile saline injected to confirm that the needle is lying freely within the joint capsule. If there is obstruction to the flow of saline and the needle is not blocked, it would indicate that the tip of the needle is embedded in a structure such as cartilage or the labrum, in which case it should be gently withdrawn or repositioned until free flow of saline is possible. If, on the other hand, on injecting saline there is no back-flow, it usually indicates that the needle is lying extra-articularly. Contrast medium should not be injected until it is ascertained that the needle is properly positioned. If the injection is given under local anaesthesia, it should be painless, although sometimes there is a sensation of distension of the joint. If pain is experienced by the patient, it indicates that the injection is being given extra-articularly and it should be stopped forthwith.

The contrast medium is absorbed rapidly from the joint itself and also from the soft tissue about the joint which results from extravasation during the manipulation or from an actual false injection.

It will be noted that there are no major vessels or nerves at the site where the needle is inserted; therefore there is no danger of damaging any delicate structures. The technique would be more difficult in gross dislocation, pelvic deformity or if there has been previous surgery on the hip. In such cases there may be distortion of the anatomical structures and the femoral pulsations may be difficult to feel. In one such case, the femoral artery was accidentally punctured, but it was possible to continue with the examination after applying digital pressure for a few minutes.

The examination should be free of any clinically significant complications if due care is taken. No articular reaction following the examination was noted; nor are we aware of any cases where avascular changes of the femoral head resulted from arthrography. These confirm previous observations that this investigation is harmless (Leveuf and Bertrand, 1937; Heublein *et al.*, 1952; Mitchell, 1963).

The most serious complication that may result from arthrography is infection of the hip joint. That is why stress is placed on the need for a thorough aseptic technique which should include proper scrubbing up, the wearing of gloves, masks and gown and adequate skin disinfection. Such skin disinfection reduces the number of bacteria, but it is doubtful if 'sterilization' of the patient's skin can ever be complete. Considering the anatomical area and the fact that most of these patients are young children and hence incontinent, skin disinfection should aim at killing as many of the bacteria as possible. The needle, introduced through such a potentially contaminated skin, can easily transmit some of the bacteria on the skin into the joint space; a relatively avascular area in which an infection can more easily establish. Consequently a reliable method for cleansing the skin should be adopted. The area is first cleaned with Cetavlon*, and the skin then painted liberally with Hibitane in spirit†. We also put a swab on the patient's genitalia in case the Hibitane runs down and trickles into the groin and onto the genitalia and perineum.

Special precautions should be taken when arthrography is being carried out following total hip replacement; an ideal situation for the establishment of any accidentally introduced infection. In these patients not only is the skin area thoroughly prepared, but the needle is introduced through a stab wound.

Taking such precautions, we have now performed over six hundred hip arthrograms without any incidence of stiffness, irritation or infection of the joint following the examination, until recently when we had a case which developed such a complication. Details of this case are given because I find it instructive and it underlines the need of proper skin disinfection.

Case Report (Case 1)
A girl of three months of age was referred for left hip arthrogram on 29.7.75 with the provisional diagnosis of hip dysplasia. The child was admitted as a day case; when she was sent to the X-ray department it was noted that her haemoglobin was only 9 g/100 ml and that she had sticky eyes. The anaesthetist was not particularly worried and anaesthesia was induced before the napkin was removed. However, when this was removed she was found to have a nasty napkin rash. Since the child was already anaesthetised, it was decided to proceed with the investigation. Only Cetavlon was used in preparing the skin as the Hibitane in spirit was considered to be too irritant to the skin with such a rash. The examination confirmed the diagnosis showing primary instability. The child was seen in the evening after the investigation, and on examination the left hip appeared satisfactory and the child was allowed to go home with an appointment to see the Orthopaedic Surgeon as an out-patient.

Two days after the examination (31.7.75) the mother reported that the child had been irritable and had vomited several times during the day and had been crying all day. The child was re-admitted when she was found unwilling to move the left hip, which was rather hot and definitely tender. The child had a temperature of 39·8 °C. Septic arthritis was suspected and aspiration of the left hip was carried out in the operating theatre when 1·5 cc of thin pus was aspirated. On examination this was found to contain numerous pus cells and a heavy growth of *Staphylococcus aureus*. A blood culture was also carried out from which *Staphylococcus* was isolated.

In spite of antibiotics, the pyrexia persisted; consequently the left hip joint was opened and drained. At operation the hip capsule was found to be lax and baggy containing pus. The antibiotics were continued, and the child recovered and is being followed up as an out-patient.

This must have been a case of septic arthritis following arthrography, presumably introduced from the skin infection. On reflection the child was not in a fit state (the haemoglobin level and the sticky eyes) for such an investigation, and secondly Cetavlon by itself is not enough to prepare the skin. The moral is:

(1) The operator must examine the patient himself, before anaesthesia is induced, and if possible before the patient is sent to the X-ray department, to ascertain that there is no skin infection or any other possible contra-indication for the examination.
(2) Skin preparation must be thorough which should include painting the area with Hibitane/spirit or with iodine solution. If such solutions cannot be applied on account of local lesions like a severe rash, then the investigation should be postponed.

4.2   Contra-indications

Usually there is no urgency for such an examination and it should be postponed if there is any evidence of:

(1) Severe skin rash in the groin.
(1) Local sepsis.
(3) Acute septic arthritis.
(4) Cold or pyrexia.

Allergy to contrast media should be excluded. Although incidence of severe allergic reaction in this investigation are extremely rare, if there is any history of allergy, asthma or sensitivity to iodine, the clinical indication for the examination should be re-assessed with the clinician responsible for the case, and the value of the information to be gained is set against the potential hazard. If after such a discussion the examination is still requested, this should be carried out with all possible precautions for preventing and dealing with any reaction. The reader is referred to the chapter 'Emergencies in the X-ray Department', Saxton and Strickland (1964).

References

Leveuf, J. and Bertrand, P. (1937), L'arthrographie dans la luxation congenitale de la hauche. *Presse Med.*, **45**, 437–440.

Heublein, G. W., Greeme, G. S. and Comforti, V. P. (1952), Hip joint arthrography. *American Journal of Roentgenology*, **68**, 736–748.

Mitchell, G. P. (1963), Arthrography in congenital displacement of the hip. *Journal of Bone and Joint Surgery*, **45–B**, 88–95.

Saxton, H. M. and Strickland, B. (1964), Emergencies in the X-ray department, *Practical Procedures in Diagnostic Radiology*, pp. 13–25, H. K. Lewis & Co., London.

* Cetavlon = cetrimide 1% in water.
† Hibitane/spirit = chlorhexidine 0·5% in 75% spirit.

5 Equipment

5.1 Radiological equipment

Although a detailed discussion of the radiological equipment is outside the scope of this book, it is felt that some comments should be made on certain important items of particular interest concerning equipment necessary to undertake such an investigation. Although arthrography can be carried out in any conventional fluoroscopy room, it is considered that the technique could be most informative and least hazardous if it were carried out in a radiographic suite as shown schematically in Figure 13.

5.1.1 *Image-intensification with closed-circuit television*

In the past a blind percutaneous approach to the joint was used. Fluoroscopy has made the examination more successful; the high rate of extra-capsular injections of contrast material has been considerably reduced. An image intensifier is essential for fluoroscopy; this linked with a closed-circuit television not only makes the technique easier but it also minimizes the irradiation to patient and staff. The basic advantage of such an image-amplified television system is the ability of the operator to place the needle accurately within the capsule at the site of preference, avoiding any damage to the growing bone or capsular structures by needling and ensuring that the injection is intra-capsular.

We use a 9/5 image intensifier – such a system has the added facility of electronic magnification if it is ever required. However, it must be remembered that such a magnification involves a considerable increase in the radiation dose and it should not be used unnecessarily. It is doubtful if magnification is ever needed in such investigations.

Of the three different types of television camera that are available, both the Vidicon and Plumbicon tubes may be considered adequate for such an investigation. The Plumbicon has a much faster response (shorter lag) than the Vidicon. Image retentivity could be troublesome during the examination especially if manipulation of the joint is carried out. However, the latest Vidicon tubes seem to have less movement lag and are comparable in this respect to the Plumbicon; this together

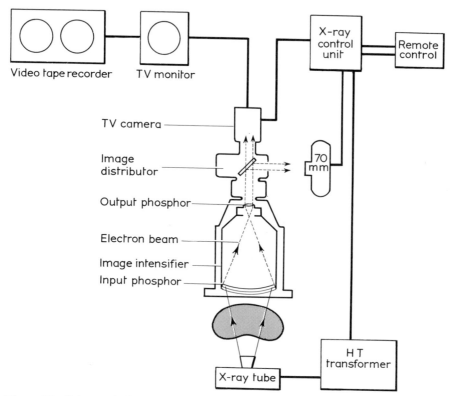

Figure 13   Schematic line drawing for a radiological suite suitable for arthrography.

with the fact that Vidicon has much less quantum mottle, i.e. 'noise', might make the Vidicon preferable to the Plumbicon considering the higher cost of the latter.

5.1.2   *X-ray tubes*

The focus size of the X-ray tube partly determines the degree of definition produced in the radiographic image. The finer the focus the sharper is the image. In the final analysis, the purpose of radiography is to bring before the observer all the information that is available; this is especially the case in such an investigation as the anatomical complexity of the young hip joint. The definition of the structures depends on the sharpness of the image. The possibility of using focal spots of smaller size than 1·0 mm and 2·0 mm combination, which is in almost universal use, should be considered. The size of a focal spot of 0·6 mm is only a third of that of 1·0 mm; this gives a small focus with acceptable rating. Radiography with the undercouch tube also has the added disadvantage of a poor focus : patient :: patient : image intensifier ratio, which distracts from good definition. We used an undercouch tube with a focal spot of 0·6 mm and we are satisfied with the degree of sharpness and definition of the image obtained.

5.1.3   *Vinten camera*

In order that a high definition image on the image intensifier output phosphor can be adequately recorded, 70 mm and 100 mm miniature fluorography has been developed.

70 mm cameras have now been in use for some time; these are capable of operating up to 6 frames per second and have a magazine which will hold up to 45 metres of polyester-based film giving approximately 600 frames. Such a camera operates without a conventional shutter; film exposure being controlled by the length of time X-rays are emitted from the tube. A mirror in the image distributor turns during the X-ray set preparation time to reflect the radiographic image into the film. After exposure, the mirror is returned and light cannot therefore enter the spot film camera during fluoroscopy.

More recently 100 mm cameras were introduced; their maximum repetition frequency is two frames per second with a proportional increase in dose. Such cameras have the added advantage of simultaneous screening facilities.

We only have experience of the 70 mm camera; we used this in some of our cases but found the size of the reproduction to be too small, and some sort of magnification, e.g. Heliocontrastor (Olde Delft), was considered helpful (Fig. 14). We have now given up using fluorography to record the findings in hip arthrography.

5.1.4   *Video-tape*

In a remarkably short time video-tape recording has won an indisputable place for itself in the field of modern television fluoroscopy. There are now completely self-contained transistorized units, recording the full television image and most offer an acceptable image quality. The video-tape recorder can be connected to any television monitor, irrespective of the television system the image intensifier uses.

The image recorded on the tape during television fluoroscopy can be played back immediately and at any subsequent time with full preservation of the diagnostic image quality.

Figure 14   opposite   (a) Contact prints from five frames of a 70 mm fluorography film in various manipulative positions, showing that when thrust was applied (second frame) there was subluxation of the head of the femur. There was also abnormal 'pooling' of the contrast medium. This was a case of primary instability. (b) Contact prints of four frames of a 70 mm fluorography series, showing a filling defect between the femoral head and the lateral part of the acetabulum due to an inturned limbus. Note that this defect is present in all positions.

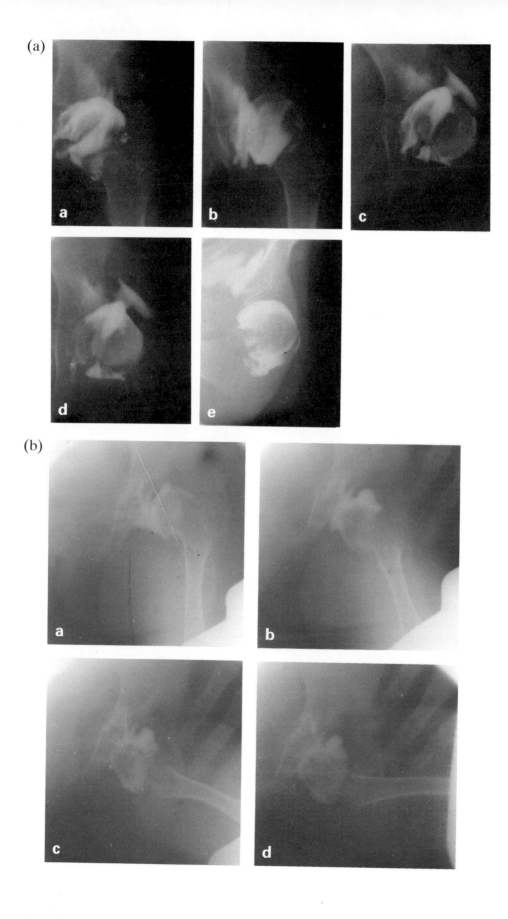

Most recorders are now capable of slow motion and even 'still' picture facilities. Inherent in such capabilities is the quality loss in the picture definition, although such a 'still' picture of an arthrogram is adequate for the radiologist to pinpoint any specific phase in the examination and to study the sequence of events in slow motion.

We found the one inch tape very satisfactory; this represents a considerable saving when its cost is compared with that of the two inch tape.

5.1.5   Remote control console

Since all table movements on the more sophisticated units are motorized, it is possible for the unit to be operated by remote control from a protective cubicle. Such a remote control unit duplicates all controls on the table in a similar manner to the control assembly mounted on the screen unit.

Remotely controlled tables are now rapidly gaining popularity. We found ours of particular help in such examinations. The radiologist can perform the investigation single-handed, fluoroscopy being controlled remotely by the radiographer from behind the protective screen. The main advantages are:

(a) No need for an assistant to screen for the radiologist, who is scrubbed-up.
(b) It quickens the examination.
(c) Both (a) and (b) help to cut down on the radiation hazard to patient and staff.

Obviously it would be uneconomical to have such a radiographic suite reserved solely for a particular investigation like arthrography; so the equipment has to be suitable for other radiological investigations. Selecting the best equipment to meet the requirements has become almost an art when you consider the overwhelming proliferation and choice of the equipment now available; besides, although the equipment is up-to-date at the time of planning, by the time such a radiographic suite becomes operational, most probably further developments and improvements have been carried out which render such units somewhat obsolete. This is due to the administrative, planning and building delays, and it is usually expected that a decision on the choice of equipment will be made years before it will be finally used. One must therefore look ahead when such a diagnostic unit is being planned. There are areas where improvement should be expected:

(1) Image intensifiers are caught up in a continuing effort to improve on the fluoroscopic image while maintaining a safe radiation level – hence the caesium-iodide image-intensifier tube. The CsI image-intensifier tube absorbs 65% of the incident X-radiation in the entry screen and converts it into image-producing information. With the conventional, or zinc-cadmium sulphide, image-intensifier tube such absorption is only 30%. In other words, these 'new generation' tubes are more than twice as efficient as the older conventional ones. The advantages of image intensifiers in radiodiagnosis are well known.

These new tubes enhance and extend these advantages mainly by improving on the detail of the fluoroscopic image. The increased resolution is distributed evenly over the whole image field. Such a uniform image quality over the total input diameter causes no variation from the centre to the periphery. Another important feature is a noticeable reduction in the X-ray quantum noise. These tubes also permit high amplification.

(2) Video-tapes, because of their convenience and versatility, will continue to play a major role in modern television fluoroscopy. Although the image quality reproduced by most of the available video-tape recorders is acceptable, it is felt that there is still room for improvement, especially when it comes to slow motion or still picture facilities, where more stability and less quality loss would improve their performance. The Video disc, which is in advanced stage of development, gives promise of considerable advancement in recording the image for communication and documentation.

5.2   Non-radiological equipment

5.2.1   *Pelvis holder for small children*

During hip arthrography on infants and young children, where manipulation on the anaesthetized child was required, it was easy to pull the child out of the field of vision during such manipulation. It was found that it would be helpful to have adequate fixation of the child's pelvis during the examination. A pelvis holder would prevent such movements; the child would be firmly held by the pelvis and hip manipulation could be carried out uninterruptedly. Such a harness can be constructed in the splint department of the hospital (Grech, 1972). The one that we use (Fig. 15) consists of two components; (i) a stainless steel bridge across the radiographic table; (ii) the saddle part, mounted on the bridge on stainless steel bars. Both the bridge and the mountings must be above the hip area so that they do not obscure the hip image. The pelvis is mainly held by the saddle which supports the lumbar spine and the sacrum.

One should have interchangeable sizes of the saddle to suit the size and age of the child. We use two sizes: a small one for children under two years of age and a larger one for older children. Those over seven or eight years do not need any fixation.

The saddle part is made from the cast of an infant's pelvis. Such a cast is made in the usual way by encasing the pelvis with plaster bandages which are allowed to dry before being removed. The pelvic girdle is renovated at the outer edges and one end covered over with plaster slab to give a perfect mould of the infant's pelvis. The inside of the mould is greased with vaseline and plaster of Paris 'cream' is poured into the mould. This is then allowed to set overnight. When the cast is set and dry, the mould is removed and the pelvis cast is ready for the splint.

From such a cast a saddle is made from a polythene sheet. This is heated in an infrared oven and moulded to fit the cast. We attached shoulder straps of leather and

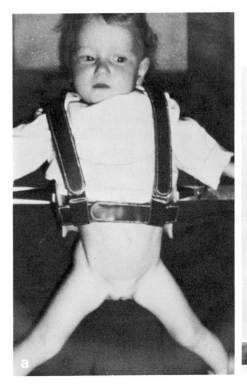

Figure 15   (a) Child firmly held by the harness while the hips are manipulated. (b) The two components of such a harness: (i) the saddle part with straps of leather and 'Velcro' mounted on stainless steel bars and (ii) the stainless steel bridge across the radiographic table.

'Velcro' to use as adjustable fastenings. Such a saddle is then mounted on stainless steel bars which in turn are attached to a stainless steel bridge across the X-ray table.

Such a harness can also be used for straight radiography of the hips.

The most important advantages that can be gained from the use of such a 'holder' are:

(a) The standard of radiography is improved;
(b) 'Coning down' is made easier;
(c) Screening time is reduced and consequently the radiation dose to the child and staff is reduced;
(d) There is no need for assistants to hold the child during the manipulation and hence the irradiation of others is limited.

References

Grech, P. (1972), A child's pelvis holder, *Radiography*, **38**, 160–161.

6 Recording of the radio-logical findings

With such radiological equipment as described in Chapter 5, the findings may be recorded by one of the following methods:

(a) conventional radiography;
(b) 70 mm fluorography;
(c) video-tape;
(d) a combination of methods (a) and (c).

Conventional radiography
Spot films represent only the static position at the time of exposure and do not show the dynamics of the joint. Reproducing all the positions of the manipulation at arthrography in a case of hip dysplasia would entail a high amount of radiation. In other conditions where one is not concerned about the joint dynamics, fewer spot films are needed.

70 mm fluorography
The camera is run at single exposure using the same mA and KV factors as for conventional radiography, but the time factor is considerably decreased. This leads to an appreciable reduction in the radiation dose. It has the added advantage of simultaneous screening facilities. The main drawback is the small size of the reproduction. It is not considered suitable for such examinations on adults; if used on infants and small children, some sort of magnification, e.g. Heliocontrastor (Olde Delft) has been found helpful.

Video-recording
The use of video-tape recorder fed from the television system enables the examination to be fully recorded. The anatomy and the dynamics across the joint are fully reproduced. It also has the advantage of being immediately available for replay. Such a tape can be played back and re-examined by the radiologist and clinician as many times as required without involving any further radiation exposure. The clinician can see the whole process as seen by the radiologist during the investigation. The case can therefore be thoroughly discussed.

Combination of video-tape and spot films
The only drawback of video-tape recording is the lack of a permanent record, unless the tape is retained which would be expensive. On the other hand such video-recording can be combined with one or two spot radiographs. These can be selected after a replay of the tape to choose the best positions to show the malapposition or abnormality. These radiographs will serve as a permanent record while the tape, once it has been examined and used in the discussion of the case, can be erased and used again.

The most suitable method depends on the case that is being investigated. If the examination is carried out to exclude hip dysplasia, then not only the static position but also the dynamics of the joint should be recorded. It is considered that in such a case the best method would be video-tape recording with one or two spot radiographs. If, on the other hand, one is not concerned with the dynamics across the joint but mainly interested in the osseous components, then appropriate radiographs will suffice.

7 Indications

Hip arthrography is often valuable in helping to solve certain orthopaedic problems, especially in children. It is not intended to be used as a routine procedure, but is limited to specifically selected cases in which it has been found to be of help to the clinician in making a decision about the diagnosis and the line of treatment. The possible indications are summarized in Table 2.

Table 2 Indications for hip arthrography

Main indications

1. Hip Dysplasia ('Congenital dislocation of the hip').

2. Dislocations associated with persistent generalized joint hypermobility.

3. Paralytic dislocations of the hip.

4. Legg–Calvé–Perthes disease.

5. Following total hip replacement.

Rarer indications

1. Following trauma or localization of loose bodies.

2. Following infective arthritis of the hip.

The majority (just over 80%) of our cases were referred with the provisional or definite diagnosis of 'congenital dislocation of the hip'. Following the pioneer work in Sweden and Italy, this condition is being diagnosed in the neonatal period more and more frequently; but in spite of all the propaganda which it received over the last twenty years or so, late diagnoses or 'missed cases' continue to occur. Children aged one year or more are still being referred to orthopaedic clinics or X-ray department with unstable or dislocated hips. Considering that neonatal management is so simple and satisfactory compared to the later treatment, this is a disturbing state of affairs and it is felt that every effort should be made to achieve an early diagnosis, although it is not certain that such neonatal dislocations and the late-diagnoses are of the same aetiology.

In a family study of neonatal (diagnosed during the first four weeks of life) and late (diagnosed after the first four weeks, usually months later) congenital dislocation of the hip, Wynne-Davies (1970a) presented evidence indicating that two gene systems are acting: one relates to dysplasia of the acetabulum and is polygenic and the other relates to the laxity of the capsule around the hip joint and is probably dominant. Both these traits may be present separately or together but in a review of 589 such patients and their families (1970b) she came to the conclusion that acetabular dysplasia which is inherited as a multiple gene system is predominantly present in the late cases while joint laxity is responsible for a high proportion of the neonatal cases.

7.1   Main indications

7.1.1   *Hip dysplasia*

The unreliability of plain radiography in the diagnosis of hip dysplasia in infants during the first few months of life is well known. (Smaill, 1968; Emneus, 1968). The case against plain radiography in such cases has been succinctly and clearly summarized by Deutschländer (1937): 'Roentgen study is of prime importance in critical evaluation of the relation of the capital epiphysis to the acetabulum. Unfortunately, this medium falls short in young children due to the fact that the upper portions of the acetabulum are cartilaginous and generally late in ossification. This state of affairs often leads to faulty interpretations and reports of satisfactory reductions are rendered when actually the reductions are poor'.

In most instances such a hip is only unstable; no true dislocation has yet occurred. Also, the ossific centre of the femoral head does not appear before the fifth month of life and sometimes even later. Consequently radiography has little to offer during these first few months of life. By the age of six months radiography is more reliable (Editorial, Journal of Bone and Joint Surgery, 1968) by which time the capital ossific centres are present. Also by this time the muscles have increased in size and strength and their pull may draw the head of the femur across the soft acetabular rim to show the displacement of the head. Actual dislocation is often gradual and may not be complete until the child begins to stand up or walk.

As has been explained in chapter three conventional radiography shows only the static position of the hip joint at the time of exposure; unless a true dislocation is present, which is not common in the early stage, it may be misleading to rely on such static data to exclude what is essentially a dynamic disturbance. This dynamic disturbance is the result of several factors; it mainly involves the development of the acetabular roof, the state of the joint capsule and the tone of the muscles acting on the joint. Therefore, it is important that plain radiography when used in the diagnosis of hip dysplasia, should include the *stress examination* if maldevelopment or malapposition is clinically suspected: that is the radiograph should be taken in a position which enhances such a malalignment. Two views ought to be included in such examination:

(1) a straight antero-posterior supine radiograph of the pelvis with the legs fully extended, pushing the knees down onto the X-ray couch, i.e. hyper-extension of the hips.
(2) Von Rosen technique – an antero-posterior view of the pelvis, with legs abducted to forty-five degrees and maximally internally rotated.

Conversely, a 'frog' view is of no use to exclude malposition, as this is the position which favours good appositional thrust across the hip joint. Its value is only to ensure good alignment following reduction.

There are other pitfalls which one should be aware of in the interpretation of the plain hip radiograph on infants (Table 3).

Table 3   Pitfalls of plain radiography of hip joint in infants

---

1. Lag in appearance of one capital ossific centre.

2. Acentral capital ossific centre.

3. Unusual appearance of the capital ossific centre – triangular or multifocal.

4. Obliquity of the pelvis – distortion of geometric data.

Lag in the appearance of one capital ossific centre
Little significance should be attached to the late appearance of the capital ossific centre, unless one side lags significantly behind the others. We have come across several cases where one ossific centre showed delayed appearance or maturity, yet clinical examination and contrast radiography revealed a normal and stable hip. The most striking example is given in this case report.

Case Report (Case 2)
This illustrative case was that of a boy who was born at home and both the midwife and the paediatrician thought that they felt a 'click' in the left hip. He was kept under observation and no further clicks were reported; clinically the hip felt stable and the child was developing normally. However, when the child was six months old, he was referred to the X-ray department by the general practitioner for a radiograph of the pelvis. This showed absence of the left capital ossific centre and was reported as '*hypoplasia of the left capital epiphysis suggestive of C.D.H.*'. The child was referred to an orthopaedic surgeon who put him in a plaster spica in a frog position. The X-ray was repeated (Fig. 16, a) when the child was nearly nine months old and again there was no evidence of the left capital ossific centre. In view of this, left hip arthrogram (Fig. 16, b) was requested and this showed a normal hip joint.

It was interesting to note that a later radiograph, when the child was eleven months, showed that the left capital ossific centre had appeared and that it had almost caught up in size with that of the right hip by the age of sixteen months (Fig. 16, c).

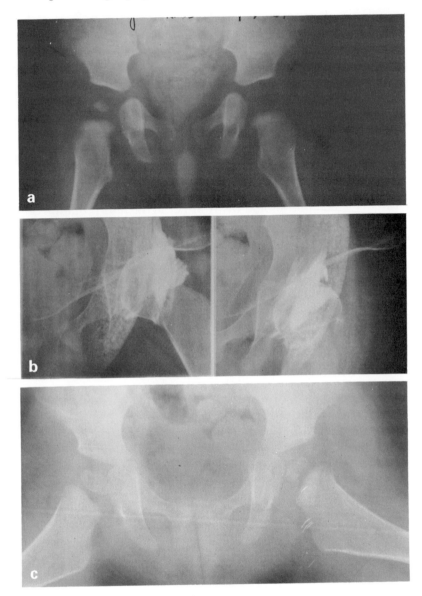

Figure 16   Case 2   (a) Antero-posterior view of the pelvis of a boy of nine months, showing absence of the left capital epiphysis. (b) Two views of left hip arthrogram showing a normal joint. (c) Repeat films at the age of sixteen months showed that left capital ossific centre had not only appeared but it was catching up in size with the other.

The eccentric capital ossific centre

Likewise, on two occasions we saw the capital ossific centre developing acentrally

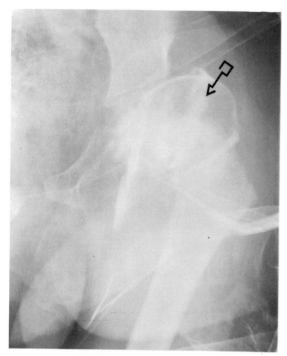

Figure 17   Right hip arthrogram on a six-month-old girl, showing the hip to be normal; but note the acentral position of the ossific centre (arrow).

within the femoral head on one side. This gave the false impression suggesting 'stand off' of the hip, yet arthrography showed a normal well-seated femoral head (Fig. 17).

Unusual shape of the capital ossific centre
As a rule the capital ossific centre appears as a spherical opacity in the middle of the cartilaginous head. Abnormal shape of this centre is usually of significance, indicating deformity of the femoral epiphysis which often presents a triangular outline due to damage from pressure by an inverted limbus, which causes flattening of the lateral aspect of the epiphysis (Fig. 18). It is claimed that rarely the ossific centre may have an anomalous appearance that is of no pathological significance. The nearest we came across such a case was that of a girl who had an adductor tenotomy on her right hip when nine months old because of a dysplasic right hip. A radiograph of the hip (Fig. 19, a) when sixteen months old showed a triangular outline and this was thought to be due to damage from an inverted limbus, but arthrography (Fig. 19, b) showed a spherical and smooth femoral head and nothing grossly abnormal was demonstrated. However the 'thorn' projection was never obviously demonstrated; also there is a suggestion of some pooling of the contrast medium in the lower joint. These features suggest that there was some pressure on the limbus which appears to be flattened and this might have had an effect on the

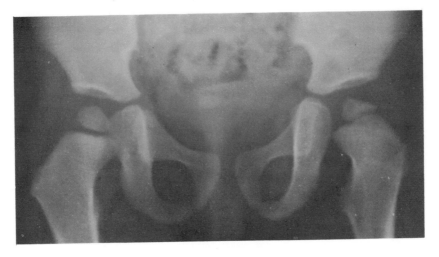

Figure 18   An antero-posterior view of the pelvis of a girl twenty-one months old who was treated by splintage for 'congenital dislocation of the left hip'. This is a good example of an epiphysis damaged by an inverted limbus which has caused flattening of the lateral part of the epiphysis. There is also obvious anteversion.

appearance of the ossific centre. However by the time she was five and a half years old, she was symptom-free and both hips looked normal (Fig. 19, c).

In another case in this series, the femoral epiphysis presented as a multifocal centre (Fig. 20); yet arthrography and follow up showed a normal hip joint.

We have come across a case which proved to be not only interesting but also instructive; the abnormal findings were not confined to the epiphysis but also to the epiphyseal plate and the metaphysis.

Case Report (Case 3)

A girl, who was delivered normally, was found to have bilateral tight clicking hips when she was eleven days old; when she was one month old she was put in a Japanese frame; she was kept on this frame until she was three months three weeks old, when both hips were found to abduct fully and easily and felt stable. As a result the frame was discontinued but the mother was advised that the child should have double nappies.

Both hips were radiographed when she was nearly six months old (Fig. 21, a); both capital epiphyses had just appeared and both hips were considered to be normal. She was followed up as an outpatient. She continued to develop normally and it was noted that she could walk with the aid of a baby walker when she was ten months old.

The pelvis was again X-rayed when she was almost one year old (Fig. 21, b). This showed apparent fragmentation of the right capital epiphysis and also irregularity and sclerosis along the metaphysis. On the left side there was possibly some widening of the epiphyseal plate and the ossific centre itself appeared blurred. No other physical abnormality was present and the child was otherwise normal. Bilateral arthrographic

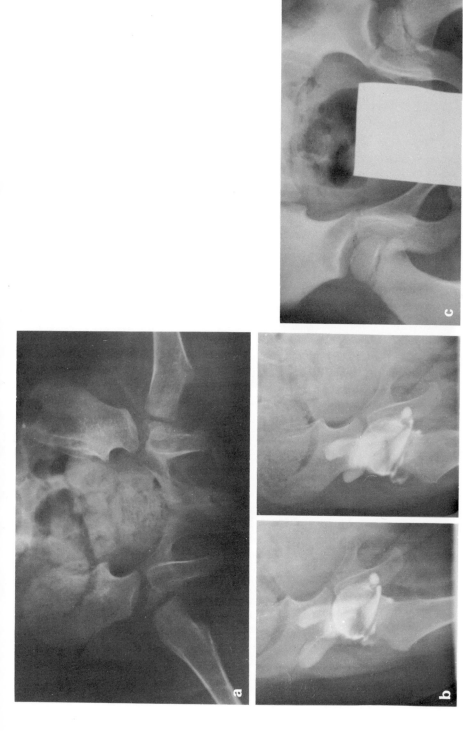

Figure 19   This girl had an adductor tenotomy on the right hip at the age of nine months on account of a dysplasic right hip. (a) Radiograph of the pelvis at the age of sixteen months showing a triangular shape of the right ossific centre. (b) Right arthrogram did not demonstrate anything obviously abnormal; there is, however, a suggestion of slight pooling of the contrast in the lower joint and also the 'thorn' projection is not clearly identifiable. These suggest some pressure on the limbus which has been almost completely flattened and this might have had some effect on the epiphysis. (c) Antero-posterior view of the pelvis at the age of five and a half years of age, showing normal hips.

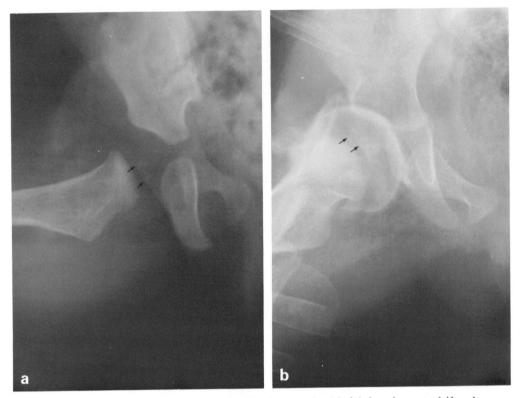

Figure 20   (a) 'Frog' view of the right hip of a six-month-old girl showing a multifocal ossific centre. (b) Right hip arthrogram shows a normal hip. This is a late film, when most of the contrast is already absorbed, so that the two centres of ossification can be seen (arrows).

studies (Fig. 21, c) were carried out at about this time. Both hip joint capsules were shown to be normal but the fragmentation and irregularity of the right metaphyseal-epiphyseal junction was again obvious; the left arthrogram appeared normal.

She was followed up as an outpatient; recently, at the age of five years, she was referred back with limping on the right side. On examination she was found to have half-an-inch shortening of the right leg and there was some limitation of movement at the hip. Her recent radiograph (Fig. 21, d) shows that there is now some reduction in the angle between the femoral neck and shaft on the left side. These angles measure 125° on the right and 115° on the left. The irregularity at the right metaphyseal-epiphyseal junction is still apparent but no fragmentation is now seen. The epiphyseal plate on the left side is now almost vertical; both femoral heads are somewhat hypoplastic and the femoral necks rather widened; but whereas the capital epiphyses are now developing reasonably normally, the metaphyseal abnormality persists.

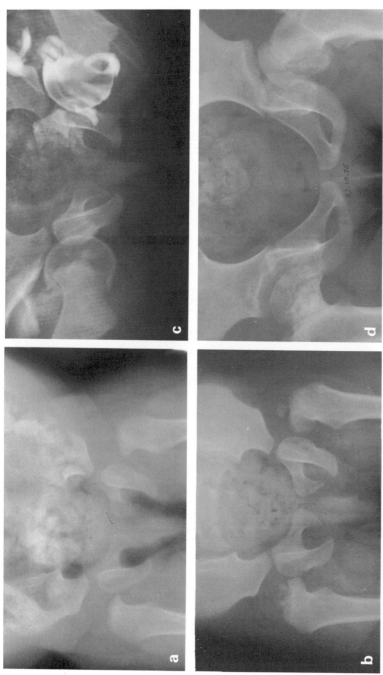

Figure 21   Case 3   Damage to the epiphyseal growth plate. (a) Antero-posterior view of a child nearly six months of age showing beginning of ossification of both capital epiphyses and both hips were considered normal. (b) Repeat X-ray of pelvis at the age of one year showing fragmentation and irregularity at the right metaphyseal-epiphyseal junction; the left hip shows some blurring of the capital epiphysis with some widening of the epiphyseal plate. (c) Bilateral hip arthrogram in 'frog' position showing both femoral heads to be smooth and well-seated and both joints to be normal. Note the apparent fragmentation at the right metaphyseal-epiphyseal junction with a fragment on the medial aspect. No such fragmentation is shown on the left side. (d) Antero-posterior view at the age of 5 years showing both femoral heads are hypoplastic and the epiphyseal plate on the left side is vertical. There is reduction in the angle between the femoral neck and shaft on the left side (115°), and both necks are widened. Irregularity at the right metaphyseal-epiphyseal junction is still present but there is now no fragmentation. Such a sequence of events is typical of that resulting from damage to the epiphyseal growth plate.

This case stimulated a lot of discussion and the differential diagnosis seems to rest between bilateral idiopathic coxa vara and epiphyseal growth plate damage. There was no clinical evidence of rickets, scurvy or any other systemic disease.

Review of the radiographs shows that in the right hip there was some sclerosis along the metaphysis with a suggestion of early 'notching' on the first film; by the time the girl was one year old both ossific centres were abnormal: the right was irregular and fragmented while the left was blurred and spotty. The 'notching' on the metaphysis on the right side is now obvious while on the left there is some widening of the epiphyseal plate. Arthrograms show that both hip joints are otherwise normal.

The femoral neck angle on the right side is within normal limits (125°) but on the left side it is slightly reduced (115°). The radiological appearances and the sequence of events are typical of those resulting from damage to the epiphyseal growth plate. With idiopathic coxa vara one would expect definite reduction of the angle between the femoral neck and shaft; obvious elevation of the trochanter giving more marked shortening of the femur. The irregularity of the metaphysis is due to 'notching' and sclerosis rather than fragmentation on the medial side of the metaphysis which one would expect to find in idiopathic coxa vara of childhood.

This is one of the complications resulting from the treatment of congenital dislocation of the hip. Mitchell (1972) has shown that reduction by abduction, in the presence of an inverted limbus in the early months of life, may cause such changes.

The upper femoral epiphysis is a pressure type of epiphysis where the nutrient vessels enter the epiphysis indirectly from the epiphyseal side. Such chronic stress, resulting from the abduction appliance, on such an immature skeleton is responsible for this epiphyseal trauma leading to impairment of the blood supply presenting radiological appearances not unlike those of osteochondritis. Such abnormal radiological findings can be seen in two vulnerable areas – the ossific centre itself and the epiphyseal plate. But as Mitchell (1972) has shown the ossific centre has got good powers of recovery, and by the time this girl was five years old, both ossific centres were developing normally while the damage to the epiphyseal growth plate leads to deformity of the metaphysis showing some degree of coxa vara together with irregularity and widening of the neck; also consequent on the blood supply impairment there could be premature closure of the epiphyseal plate resulting in short widened femoral necks with some degree of coxa vara, as can be seen in this case.

It might be opportune to remind ourselves about the possibility of such iatrogenic damage to the epiphyseal plate which might result from stress following maximal abduction or forceful manipulation. Forceful abduction in the old-fashioned 'full frog position' may cause damage to the femoral epiphysis even in the absence of an inverted limbus.

Obliquity of the pelvis
Numerous geometric lines and acetabular angles have been proposed to evaluate

abnormalities across the hip joint. As Wittenberg (1964) explained, most of these have severe limitations because they rely on anatomical landmarks unrelated to the hip joint itself, but lying in different coronal planes and, therefore, liable to geometric distortion from rotation of the pelvis on such infants (Fig. 22). This is why

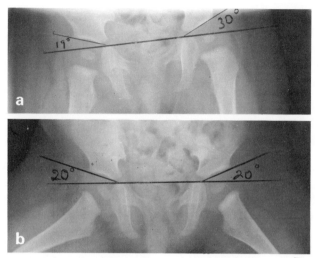

Figure 22 Two views of the pelvis of a child taken on the same day. The top one is 'at rest' position; the slight obliquity of the pelvis is reflected in the marked difference in the acetabular angles. The bottom radiograph is the von Rosen view with the pelvis straight and the acetabular angles are shown to be equal.

we constructed the pelvis harness (Fig. 15); it helps to eliminate such obliquity of the pelvis. Not only lateral pelvic rotation can distort such geometric data, but even arching of the lumbar spine can affect the acetabular angle. Consequently, the plain radiograph which demonstrates only the osseous parts of the joint in a static state should be perfectly straight and level to be of any significant value in assessing the geometric lines and acetabular angles. Besides the possible pitfalls already described, plain radiography does not provide any information about the dynamics across the joint, the relative position of the cartilaginous head, or the fibrocartilaginous labrum and the capsule. Hip arthrography enables a more accurate study not only of the anatomy, but also of the dynamics of the hip joint.

The reasons why arthrography was requested in cases referred with the provisional or definite diagnosis of hip dysplasia are summarized in Table 4.

Doubt or difference of opinion between clinicians about the diagnosis
Uncertainty about the diagnosis or difference of opinion between clinicians, usually paediatricians and orthopaedic surgeons, seems to be a common reason for referring the child for hip arthrography. In such problematic cases, arthrography is helpful in

establishing the diagnosis and may avoid unnecessary, repeated plain radiographs. Such clinically doubtful dysplasic hips are often found to be cases of primary instability. We came across several cases of normal hip joints which were previously

Table 4   Reasons for requests for arthrography in cases of hip dysplasia

1. Doubt or difference of opinions between clinicians.

2. Failed closed reduction.

3. ? Recurrent dislocation.

4. Prior to open reduction.

5. Limited abduction syndrome.

6. Following open reduction – to ascertain a good result.

diagnosed as being dysplasic; the possibility of hip dysplasia was suggested from the appearances of the capital ossific centres on plain radiography, as listed under pitfalls of plain X-rays, whereas arthrography shows them to be normal joints.

Failed closed reduction
In about a third of the cases referred, the investigation was requested after closed reduction had failed. Arthrography will help to show the cause of the obstruction. The usual obstructive element is an inturned limbus, or hour-glass deformity of the capsule; other lesions found to be responsible for the failure of reduction were a pad of fat (Fig. 23), a cartilaginous cushion or a bulky ligamentum teres (Figs. 24 and 25) preventing the proper sitting of the femoral head.

Case Report (Case 4)
D. G., a girl, was born by caesarean section on account of placenta praevia and prolapsed cord. There was no fetal distress and the baby was in good condition at birth; she continued to progress satisfactorily. When she was nearly six months old, she was noted to have limitation of eversion of the left hip and also asymmetry of the buttock creases. Radiograph of the pelvis (Fig. 24, a) was reported as 'The capital epiphyses have not yet appeared; probably slight stand-off on the left side'. The child was put in a Von Rosen splint and re-X-rayed after two and a half months (Fig. 24, b). This showed apparent delay of appearance of the capital ossific centre and definite stand-off on the left side. The hip was manipulated and put in plaster in a frog position. When the child was thirteen months old she was again X-rayed (Fig. 25, a); this showed increased lateral displacement of the left hip and 'unusual appearance of the femoral head'. In view of this a left hip arthrogram was requested. This (Fig. 25, b) showed flattening of the whole femoral head with an apparent depression at the insertion of the ligamentum teres (△) which was particularly visible in the frog position. A rather long ligamentum teres was also suspected in' the other views (arrows). These findings suggested a dislocation of the femoral head which was not reducible. Following this examination exploration of the hip joint was decided upon. The ligamentum teres was found to be

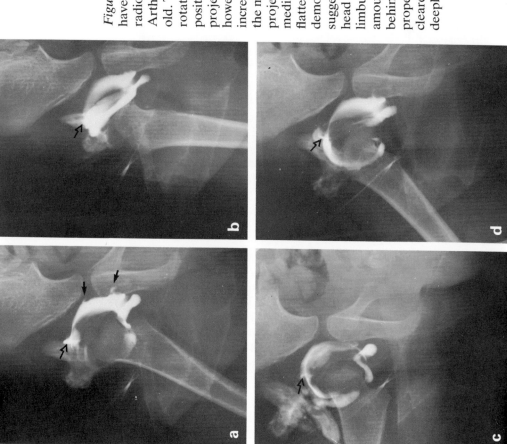

Figure 23 This girl, when five months old, was found to have 'instability of the right hip' on clinical examination. A radiograph of the right hip showed 'stand-off'.

Arthrography was carried out when she was seven months old. These four frames show the right hip in (a) external rotation, (b) neutral (c) frog and (d) full abduction positions. The open arrow points to the 'rose-thorn' projection which appears to be in normal position; however there is pooling of the contrast medium medially, increase in the gap between the floor of the acetabulum and the medial aspect of the femoral head; two small projections of contrast agent (straight arrows) on the medial aspect of the joint between which there is apparent flattening of the synovial outline; these are best demonstrated in external rotation. These appearances suggested soft tissue interposition between the femoral head and floor of acetabulum. At open exploration the limbus was found to be normal but there was excessive amount of 'thick tissue' in the floor of the acetabulum behind the ligamentum teres which was preventing the proper sitting of the femoral head. This 'thick tissue' was cleared and then the femoral head appeared to sit more deeply in the acetabulum.

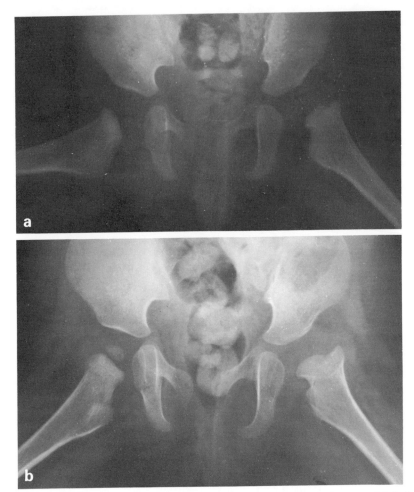

Figure 24   Case 4   Radiographs of the pelvis of a girl at the age of (a) six months and (b) nine months respectively showing delay in the appearance of the capital ossific centre and 'stand-off' on the left side.

Figure 25   (a) Antero-posterior view of the pelvis when the girl was thirteen months confirming the lateral displacement of the left hip. The femoral head was reported as presenting 'unusual appearances' as if it was squashed on to the neck of the femur. (b) The arthrograms with the line drawing confirm flattening of the femoral head and increase in the gap between the acetabulum and the head suggesting an obstructive element. In the frog position there is a depression (△) which could possibly be at the insertion of the ligamentum teres; the arrows point to a linear translucent shadow which could be due to a long ligamentum teres.

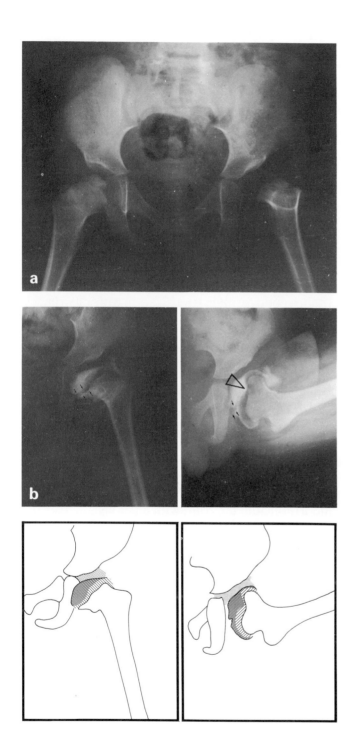

enormously bulky and it had indented the top of the femoral head which was flattened. It was also found that the best containment of the femoral head was obtained when the hip was put into slight internal rotation and abduction.

The child was then put in a plaster spica; this was renewed and kept on until she was two and a half years old. She is being followed up as an out-patient; when seen recently, at the age of almost five years, she was running about satisfactorily but with a slight limp.

Recurrent dislocation

It is not certain whether the reduction in these cases was ever complete and perhaps most of these should be included with the failed closed reduction group. The difficulty of obtaining good results in 'true' dislocation cases by closed reduction is well known; the value of arthrography in such cases was stressed by Leveuf (1948) when he concluded that 'it would certainly be better to avoid reluxations and secondary subluxations. This is why it is absolutely necessary to check the reduction by means of arthrography in all cases of congenital dislocation of the hip. Then one can verify that it is impossible to obtain a good result in true luxation without primary open reduction'.

Prior to open reduction

It is advantageous for the surgeon to know what he is going to find at operation. Arthrography should be considered in all cases where surgery is contemplated, whether open reduction or corrective osteotomy.

Case 5 illustrates the importance of this investigation in such cases. A plain radiograph of a year old girl showed 'subluxation' of the right hip. This was thought to be reduced on internal rotation. Consequently a de-rotation osteotomy of the femur was carried out. During a physical examination five months after the operation, the hip was thought to be unstable and 'recurrence of the dislocation' was suspected; the child was then referred for arthrography. When this was carried out (Fig. 26) it showed a dislocated and irreducible hip joint with an elongated capsule presenting an early hour-glass deformity.

Since arthrography was not carried out before the de-rotation osteotomy, one cannot be sure that the reduction on internal rotation was ever complete; therefore there must be grounds to question whether the de-rotation osteotomy was the proper line of management as we are not sure that it did reduce the original 'subluxation'.

Limited abduction syndrome

Sometimes one sees a child with limited abduction of the hip that is not due to 'congenital dislocation of the hip' but to a benign condition caused by tight adductors.

Arthrography will show normal appearances and therefore it will exclude hip dysplasia; consequently surgical intervention or unnecessary prolonged immobilization will be avoided.

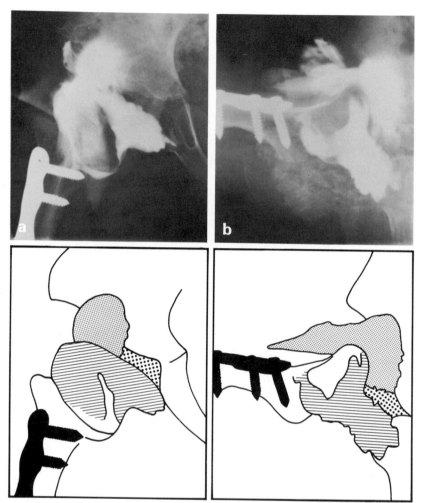

Figure 26   Case 5   A plain radiograph of this girl when she was one year old showed 'subluxation of the right hip which was thought to be reduced in internal rotation'. Consequently a de-rotation osteotomy of the right femur was carried out. Hip arthrography was carried out five months after the operation because there was 'recurrence of the dislocation' which shows a dislocated, irreducible hip joint with an elongated capsule.

Following open reduction

Arthrography is sometimes carried out to ascertain complete reduction. I do not consider that this is really necessary, and it should only be considered in those cases where satisfactory reduction is uncertain or re-dislocation suspected (Fig. 27 and 28).

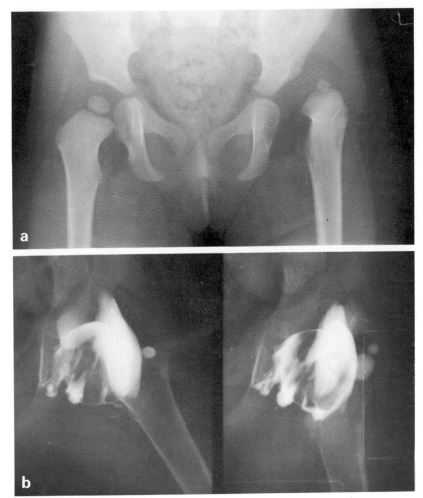

Figure 27   This girl was known to have had an unstable left hip from the age of three months, but in spite of the closed reduction and immobilization the instability persisted. (a) An antero-posterior view of the pelvis when she was ten months old showed lateral and cephalad displacement of the femoral head; small capital ossific centre and anteversion of the femoral neck. Three weeks later an adductor tenotomy was carried out followed by manipulation and immobilization in plaster of Paris. When the plaster was removed three months later, arthrography was carried out. (b) Arthrogram of left hip confirming the head to be well seated and it was shown to be stable. The fusiform shadow represents contrast in the psoas sheath.

7.1.2   *Dislocation associated with persistent, generalized joint hypermobility*

The range of joint movements varies between comparable individuals. Where the joints are unduly lax and the range of movements is in excess of the accepted norm, the subject is considered to be afflicted by 'the hypermobility syndrome' (Kirk, Ansell and Bywaters, 1967). Such excessive joint movement in three or four joint pairs is estimated to be between four and seven per cent of the population (Sutro, 1947; Carter and Wilkinson, 1964).

Capsular laxity is an important aetiological factor in the development of congenital dislocation of the hip (Massie and Howarth, 1951). In the beginning such capsular

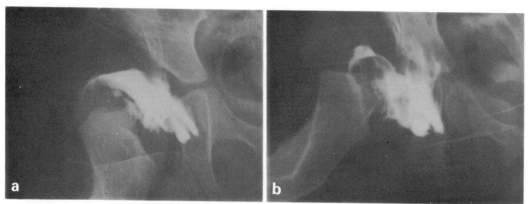

Figure 28   A boy who was found to have dislocation of the right hip at the age of six months was treated by a subtrochanteric osteotomy. When the plaster was finally removed the head was thought to be still not well-seated and unstable. Arthrography was requested for more information: (a) leg in neutral position and (b) in abduction and external rotation. Both views show the head to be dislocated and irreducible.

laxity was considered to result from mechanical stretching, but it is now thought to be a part of generalized joint laxity affecting most joints. It appears that there are two forms of generalised joint laxity which develop during pre-natal life (Carter and Wilkinson, 1964).

(a)   Temporary laxity   – usually found in girls, possibly of hormonal origin;
(b)   Persistent laxity   – which may occur in either sex and is often familial. Such generalized joint laxity is a feature of certain genetically determined connective-tissue disorders, which can present with an atypical dislocation of the hip.

The degree of hypermobility and the incidence of joint dislocations are closely related (Beighton and Horan, 1969). Although such joint hypermobility might represent one end of the clinical spectrum of normal movements (Kirk, Ansell and Bywaters, 1967), such persistent generalized joint laxity is a feature of many clinical syndromes. Among the various conditions presenting with persistent generalized hypermobility of joints, in this series, hip dislocation was noted in:

(a) Ehlers-Danlos syndrome – where the generalized capsular laxity is associated with skin laxity.
(b) Arthrochalasis multiplex congenita – a name given by Hass and Hass (1958) to the condition of generalized hypermobility of joints without obvious skin laxity, which is said to be an entity separable from Ehlers-Danlos;
(c) Down's syndrome – Penrose and Smith (1966) stated that mongols often show a generalized hypermobility of joints;
(d) Garrod-Davies syndrome – where, besides the genito-urinary abnormality, there is muscle hypotonia and congenital deficiency of the abdominal muscula-ture. Hip dislocation was noted in one case.

In these cases of persistent joint hypermobility, arthrography will confirm that dislocation does occur and that it is followed by complete reduction; it also shows the dynamics produced across the joint and the mechanics involved in producing the dislocation.

7.1.3 *Paralytic dislocations of the hip*

Paralytic subluxations or dislocations of the hip can be caused by: (a) overaction of unopposed muscles; (b) shortening of paralysed muscles, and (c) weight-bearing on a partly paralysed leg. Such acquired hip dislocations are usually associated with neuro-muscular diseases. Now that the incidence of poliomyelitis is fast decreasing, the most common are cases of paralytic scoliosis or spina bifida cystica – some of these patients develop paralytic subluxation or dislocation of the hip that will need reduction and a stabilizing procedure. Following reduction of such dislocation, the femoral head can be held in position by transferring the insertion of the psoas muscle from the lesser trochanter into the greater trochanter, bringing the muscle through an opening in the iliac bone.

Once dislocation has occurred, muscle contraction may follow together with capsular fibrosis and adaptive cartilaginous or bony changes. These may affect the operative procedure and arthrography is sometimes requested before such correc-tive and stabilizing techniques to see whether it is possible to reduce the dislocation; if not, what is preventing it and so deciding on the most suitable treatment for achieving joint stabilization.

7.1.4 *Legg–Calvé–Perthes disease*

Jonsäter (1953) carried out an extensive correlative study of arthrography of the hip and of histopathology of Legg–Calvé–Perthes disease from trephine biopsies. This and similar studies (Katz, 1968; Ozonoff, 1973) have helped us to understand the progress of the disease better. Such arthrographic studies have been found valuable in these cases to assess the shape, congruity and coverage of the femoral head.

It is not being suggested that arthrography should be carried out as a routine investigation in Perthes disease; it adds little information in the initial and latest stages of the disease but it might give helpful information in the fragmentation and reparative stages. Actually this investigation is rarely used in Perthes disease in England, but it is more widely utilized in other countries, especially in the United States; for example in Newington Children's Hospital, Connecticut, almost 40% of the hip arthrograms performed are for Perthes Disease (Ozonoff, 1975). It has also been suggested that arthrography might be of value in assessing the prognosis of a particular case and in deciding on its management. Catterall (1971) evaluated the use of arthrography in Perthes disease and he found arthrography valuable in assessing the natural history of the disease; he also came to the conclusion that the arthrographic appearances cannot be anticipated from the plain radiograph. It was found that nearly always there is some reduction in the epiphyseal height but this is frequently compensated for by the adaptive changes in the acetabulum. From the arthrographic studies, he differentiated the femoral heads into (a) spherical (b) round and (c) flattened varieties. He concluded that a hip with an arthrogram showing severe flattening of the femoral head will proceed to a poor result, while a round or spherical head will lead either to a fair or good result. Consequently the shape of the femoral head is of some importance when the need of treatment in late cases is being considered.

7.1.5 *Following total hip replacement*

Total hip replacement is now frequently and extensively used. Such an operation replaces both the acetabular socket and the head and neck of the femur by a prosthesis. There are now many types of such prostheses in use but these can be divided into two groups:

(a) *Metal-to-metal articulation* – like the McKee–Farrar prosthesis where both parts are of vitallium;
(b) *Plastic-to-metal articulation* – like the Charnley prosthesis where the socket consists of a high density polyethylene and the neck and head of stainless steel.

Both parts of the prosthesis are cemented into bone using methylmethacrylate, which is usually made radiopaque by adding to it barium sulphate.

The commonest complications which may lead to failure of this operation are loosening of the prosthesis or a low-grade infection which could lead to the former. Such complications should be suspected in patients with persistent pain following total hip replacement. The time lag of these complications is variable – from a few months to years.

Definite diagnosis of such complications is not always possible, especially in the early stages, by clinical examination or plain radiography. Arthrography has been shown to be valuable in the investigation of such complications (Salvati *et al.*, 1971). According to Wilson *et al.* (1971) the reliability of arthrography in such cases is said to be high and they conclude that 'while a normal arthrogram does not definitely rule out the presence of a complication, a positive arthrogram appears to be diagnostic'.

7.2   Rarer indications

7.2.1   *Following trauma or localization of loose bodies*

Arthrography is sometimes considered in the investigation of capsular tears which should show extravasation of the contrast medium; the investigation is also occasionally carried out following reduction of fracture dislocation of the hip, especially if there is some doubt about its stability. It has also been used in cases where a chondral fracture was suspected. Arthrography may also be used occasionally to investigate the possibility of loose bodies in a hip joint, where plain radiography is inconclusive.

7.2.2   *Infective arthritis of the hip*

Acute purulent infections of the hip joint are more common in infants and young children than in older ones. Such purulent arthritis usually develops secondary to bacteraemia from upper respiratory infections or sepsis elsewhere. The coagulase-positive staphylococcus is the commonest organism to affect the hip joint in the child. In the acute stage the hip joint is held flexed and the child resents any leg movement. Occasionally, the infection is rather mild and it may be difficult to distinguish from tuberculous infection clinically. Persistent suppuration destroys the articular cartilage and associated bone changes are common in all types of purulent arthritis in infants and children, although there is usually a time lag for such bone changes to become radiographically manifest. Not infrequently the head of the femur is destroyed. Following such joint infections, thickening of the capsule and pericapsular tissues occurs; there is accumulation of the synovial exudate in the joint capsule and as a result it distends. Severe capsular distension can lead to pathological subluxation or even frank dislocation.

Arthrocentesis might be needed to aspirate the joint to confirm the diagnosis and identify the responsible organism or even to instil medications in the joint. In the late stages of the disease such a dislocation becomes unreducible and plain radiography will not show the extent of the cartilaginous or soft tissue destruction. In such problem cases, especially if there is doubt about the original cause of the dislocation due to lack of clinical details, arthrography might have to be resorted to in order to establish the diagnosis or even prior to surgery.

References

Beighton, P. and Horan, F. (1969), Orthopaedic aspects of Ehlers–Danlos Syndrome. *Journal of Bone and Joint Surgery*, **51–B**, 444–453.

Carter, C. and Wilkinson, J. (1964), Persistent joint laxity and congenital dislocation of the hip. *Journal of Bone and Joint Surgery*, **46–B**, 40.

Catterall, A. (1971), The Natural History of Perthes' Disease. *Journal of Bone and Joint Surgery*, **53–B**, 37–53.

Deutschländer, K. (1937), Der umbau des Hüftgelenks nack der blutigen radikaloperation der angeborenen Huftverrenkung. *Ztachr. f. Orthop.*, **67**, 116–123.

Editorial (1968) *Journal of Bone and Joint Surgery*, **50–B**, 453.

Emneus, H. (1968), A note on the Ortolani–Von-Rosen–Palmen treatment of congenital dislocation of the hip. *Journal of Bone and Joint Surgery*, **50–B**, 537.

Hass, J. and Hass, R. (1958), Arthrochalasis multiplex congenita. *Journal of Bone and Joint Surgery*, **40–A**, 663.

Katz, J. F. (1968), Arthrography in Legg–Calvé–Perthes Disease. *Journal of Bone and Joint Surgery*, **50–A**, 467–472.

Kirk, J. A., Ansell, B. M. and Bywaters, E. G. L. (1967), The hypermobility syndrome. Musculoskeletal complaints associated with generalised joint hypermobility. *Annals of the Rheumatic Diseases*, **26**, 419–425.

Leveuf, J. (1948), Results of open reduction of true congenital luxation of the hip. *Journal of Bone and Joint Surgery*, **30–A**, 875–882.

Massie, W. K. and Howarth, M. B. (1951), Congenital dislocation of the hip. Part III, Pathogenesis. *Journal of Bone and Joint Surgery*, **33–A**, 190.

Mitchell, G. P. (1972), Problems in the early diagnosis and management of congenital dislocation of the hip. *Journal of Bone and Joint Surgery*, **54–B**, 4–12.

Ozonoff, M. B. (1973), Controlled arthrography of the hip. A technic of fluoroscopic and recording. *Clinical orthopaedics*, **93**, 260–264.

Ozonoff, M. B. (1975), Personal communication.

Penrose, L. S. and Smith, G. F. (1966), *Down's Anomaly*, Churchill, London.

Salvati, E. A., Freiberger, R. H. and Wilson, P. D. (1971), Arthrography for complications of total hip replacement. *Journal of Bone and Joint Surgery*, **53–A**, 701–709.

Smaill, G. B. (1958), Congenital dislocation of the hip in the newborn. *Journal of Bone and Joint Surgery*, **50–B**, 524.

Sutro, C. J. (1947), Hypermobility of bones due to 'overlengthened' capsular and ligamentous tissues: a cause for recurrent intraarticular effusions. *Surgery*, **21**, 67.

Wilson, P. D., Freiberger, R. G. and Salvati, E. A. (1971), Arthrography and total hip replacement. *Journal of Bone and Joint Surgery*, **53–A**, 1245–1246.

Wittenberg, M. H. (1964), Malposition and dislocation of the hip in children and childhood. *The Radiologic Clinics of North America*, No. 2, W. B. Saunders Company, Philadelphia and London.

Wynne-Davies, R. (1970a), A family study of neonatal and late-diagnosis congenital dislocation of the Hip. *Journal of Medical Genetics*, **7**, 315–333.

Wynne-Davies, R. (1970b), Acetabular dysplasia and familial joint laxity: two etiological factors in congenital dislocation of the hip. *Journal of Bone and Joint Surgery*, **52–B**, 704–716.

8 Interpretation of arthrograms

Doing an arthrogram is one thing, but interpreting the findings is another, and an operator must perform many hip arthrograms before he is in a position to interpret the findings with confidence. We owe our present knowledge of the interpretation of hip arthrograms to the pioneer work of Severin. This is based on his 'cinnoidin arthrograms' where he injected the hips of stillborn children at full term with a mixture of cinnabar and celloidin. When the 'cinnoidin' had solidified, the arthrographic casts were checked by radiography and they were subsequently dissected and compared with the radiographic appearances.

Interpretation of arthrograms is based on the correlation of the radiographic findings to the anatomical structures. An attempt is made in this section to give some guidelines to help such interpretation and also, whenever necessary, to accompany the illustrations by explanatory line drawings, using the same key throughout: i.e. white representing bone; shading indicating the presence of contrast medium, the heaviness of the shading depends on the amount of contrast present, and the cartilage is represented by dots.

8.1 Normal arthrogram

In the infant or young child most of the spherical femoral head is cartilaginous and consequently radiographically translucent. The injected contrast material is uniformly distributed in the joint space forming a thin layer of contrast around the head without any persistent pooling. The fibro-cartilaginous labrum is well outlined by the contrast agent (Fig. 29) and its outer margin is shown by the 'rosethorn' projection on the antero-superior aspect of the joint. The deep 'groove' on the infero-medial aspect of the arthrogram is caused by the ligamentum transversum. As the outer layer of this ligament is a direct continuation of the limbus, these two landmarks – the rose-thorn and the groove – demonstrate the extent of the fibro-cartilaginous element of the acetabulum. The cartilaginous acetabulum should cover at least half of the femoral head. There should not be pooling of contrast medium in the bottom of the acetabulum. In the normal arthrogram the ligamentum teres cannot usually be identified, although in some cases its origin can be located by two small projections of contrast medium running downwards and medially into the acetabular fossa.

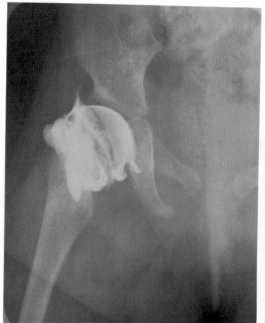 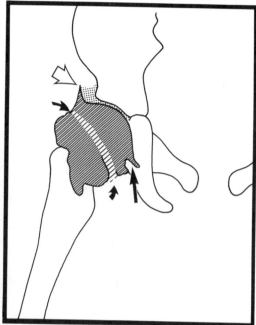

Figure 29   Normal right hip arthrogram in a seven-month-old child together with a schematic drawing. The black arrow points to the 'groove' of the ligamentum transversum. The white arrow indicates the 'rose-thorn' projection. The two short curved arrows show the 'zona orbicularis'.

The capsule completely covers the femoral head and neck and reaches down to the intertrochanteric line anteriorly, but not so far posteriorly. The capsule is constricted around the waist of the neck by the '*zona orbicularis*', forming a band where there is less contrast material as the space is taken up by a bundle of deeply placed circular muscular fibres, and this band therefore appears less radio-opaque than the rest. Such an impression of the zona orbicularis is more obvious in the young.

Within the spherical cartilaginous head one can see the epiphyseal centre of ossification. It gets bigger with age until in the adult it occupies the whole of the head except for the articular cartilaginous layer, which appears as a sharp thin translucent line underneath the rim of contrast material around the femoral head.

The movements of the femoral head within the acetabulum can be roughly described as 'concentric', i.e. they occur around an imaginary centre. Therefore there should not be any abnormal pooling of contrast medium. This layer of contrast agent around the femoral head should be approximately parallel with the acetabulum in all positions (Fig. 11); such findings would confirm that the head is normally seated within the acetabulum and is stable; it also shows congruity between the femoral head and its acetabular socket.

Jonsäter's caput-index (1953) is a calculation to determine the sphericity of the femoral head; it is represented by the ratio of the height of the femoral head to half its greatest width. To find this out you first measure the greatest width of the arthrographic outline of the femoral head (AB) (Fig. 30); you then draw a perpendicular from the centre of AB to the articular surface of the head (CD).

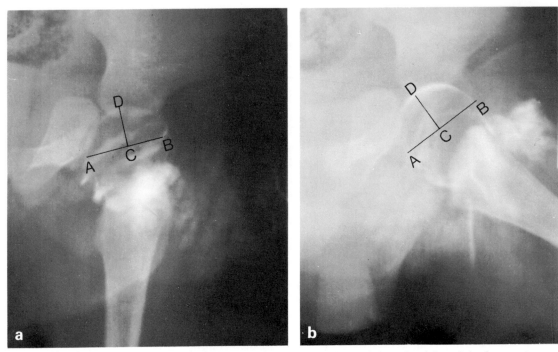

Figure 30   Caput index. This is designed to give an idea of the sphericity of the femoral head. AB is a line drawn along the greatest breadth of the femoral head as outlined by the arthrogram and it measures 30 mm on the radiograph. C represents the midpoint of AB and line CD is thrown perpendicular at C to reach the articular surface of the femoral head. CD is found to measure 15 mm. The caput index is presented by the relationship of the height of the femoral head (15 mm) to half its greatest breadth

$$\left(\frac{30}{2}\text{mm}\right) \qquad \text{or } \frac{15}{15} = 1.$$

$$\text{The caput index} = \frac{\text{CD}}{\frac{1}{2}\text{AB}}$$

In the normal femoral head this is about 1.

A fair amount of space is deliberately given to the description and illustration of the normal hip. It is often difficult for the trainee to appreciate what a normal hip joint is like as the opportunity to examine a normal hip does not occur frequently. Unless one fully understands the normal appearance, one cannot start to interpret the abnormal.

8.2   Abnormal arthrograms

Disease in the hip joint may affect either the bony parts or the soft tissue parts, but often both. Arthrography may be needed to demonstrate the full extent of the abnormal changes. In the interpretation of abnormal arthrograms special attention must be paid to the following aspects:

(a) Capsular capacity – One can get a good impression about the state of the joint from the ease and amount of fluid one can inject before there is flow-back in the syringe. The average normal joint will take only a few millilitres, while a lax capsule will take much more. Likewise if the capsule is stretched due to a dislocation, and especially if such a dislocation is associated with an hour-glass deformity of the capsule, then the capsular capacity is increased. If there is loss of the elasticity of the capsule and it is difficult to inject any fluid without flow-back, this would indicate a contracted fibrotic capsule most likely resulting from a pyogenic arthritis.

(b) Position of the two landmarks – The 'rose-thorn' projection and the 'groove' denote the site and extent of the fibro-cartilaginous labrum. These will also show the extent of the coverage of the femoral head.

(c) The joint space – The gap between the femoral head and the acetabulum should be regular in width and symmetrical. Soft tissue interposition will interfere with this 'parallelism' between the acetabular margin and the rim of contrast material around the femoral head.

(d) The articular surface – This is represented by a translucent layer on either side of the rim of dye around the head. It should be smooth and regular. Erosions or depressions in the articular layer will be demonstrated by the contrast medium and such filling defects will interfere with the regular outline of the articular surfaces.

(e) The shape of the femoral head – This should be smooth and spherical. It should be congruous with the acetabular socket. The sphericity of the head is calculated by the caput index which should be about 1. An index appreciably less than one indicates a flattened femoral head.

(f) Stability of the head – The head should be contained within the acetabular socket throughout the full range of movements. There should not be any pooling of the dye or displacement of the head from one position to the other.

(g) Range of movements – Although there may be some variation in the range of movements between comparable individuals, joint movements should be full and free. Any limitation or hypermobility should be noted, but the joint should be stable throughout the range of movements.

8.2.1   *Hip dysplasia*

I consider the term 'congenital dislocation of the hip' (or simply C.D.H.) as a misnomer. It is hoped that most of these cases will never actually dislocate. Hip dysplasia is probably a better term. *Dysplasia* is made up from two Greek words: *dys* meaning *bad* and *plassein* meaning *to form*. Therefore, dysplasia stands for abnormal development or growth and this is what really happens. Definite diagnosis of hip dysplasia in infants from the plain radiograph can only be made when there is actual dislocation of the joint. Acetabular changes usually associated with hip dysplasia, like increase in the acetabular angle, are not necessarily present; in actual fact, they are often absent.

In the dysplasic case, arthrography may show central dye pooling or soft tissue enlargement; the enlarged ligamentum teres may be visualized by a broad, relatively radiolucent line, running from the head towards the transverse ligament. The labrum may be elevated or inverted which may lead to constriction of the joint capsule and renders such dislocation non-reducible.'Finally the head may lose its spherical shape and becomes flattened.

The value of arthrography in the diagnosis of problematic cases of hip dysplasia was first shown by Severin (1939) and more recently by Mitchell (1963); Astley (1967); Felländer *et al.* (1970) and Grech (1972). It should help in the diagnosis of primary instability where plain radiography is usually of no use; it should differentiate partial displacement cases from those showing true dislocation, which is most important from the management point of view, and give useful information about the nature and extent of the soft tissue interposition, in true displacement cases.

Mitchell's classification of dysplasic hips has been found useful in describing and illustrating the arthrographic findings in this condition. He suggested three degrees of displacement:

(a) Primary instability;
(b) Partial displacement;
(c) Complete displacement which may be either of the 'tight' or 'loose' variety.

The radiographic appearances – on the plain X-ray and arthrogram – are summarized in Table 5.

Primary instability
Primary instability is a better term than 'predislocation'; detection of such cases should prevent them from progressing to dislocation. As Mitchell pointed out 'many

Table 5 Classification and summary of the radiological findings in hip dysplasia

Stage	Terminology	Synonym	Radiological findings		
			Plain radiography	Arthrography	
I	Primary instability	Predislocation	Normal appearances	Increased capacity of the joint capsule. There may be 'pooling'. The limbus is not well outlined in lateral rotation but it is well shown when the hip is abducted and femoral head slips in socket. (Fig. 31)	
II	Partial displacement	True subluxation	'Stand-off' or subluxation may be noted on 'stress' examination.	Increased capsular capacity and therefore laxity. The limbus is squashed against the roof of the acetabulum, especially on lateral rotation of the femoral head and therefore 'landmarks' disappear. However as the femoral head slips back into the socket on abduction, the limbus reappears. Subluxation is obvious on telescopy of the joint (Fig. 32).	
III	Complete displacement	(a) Tight dislocation	Some increase in the gap between the acetabulum and the head of femur but such displacement is never marked and could be mistaken for partial displacement.	The normal position of the limbus is absent; soft tissue interposition between the head and acetabulum caused by the inturned limbus is outlined (Fig. 33) as it slips behind the femoral head.	
		(b) Loose dislocation	The displacement is obvious and besides the lateral there may be also cephalad displacement. This overt dislocation may also be associated with hypoplasia of the capital ossific centre.	The normal position of the limbus is absent. There is increase of capsular capacity and the femoral head displacement is more marked (Fig. 34); the capsular infolding which is inverted with the limbus presents various degrees of hour-glass constriction (Figs. 35 and 36).	

of such hips will develop normally without treatment while others only do so if they are kept in abduction for some time'. Here the main fault appears to be increased laxity of the capsule. Plain radiography is usually normal but arthrography shows increase in the capacity of the joint capsule and pooling of the dye (Fig. 31). Due to this capsular laxity, the femoral head also has a tendency to show increased degree of 'play' within the acetabulum. As a result the labrum may not be so well outlined

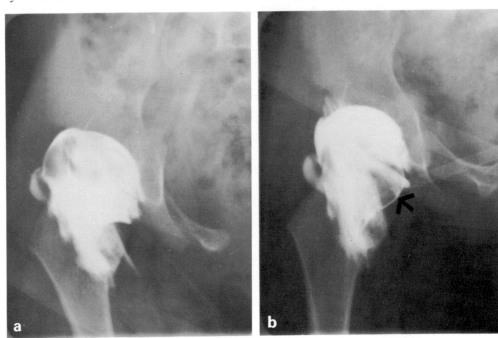

Figure 31   Primary instability of the right hip. Right arthrogram on a thirteen-month-old girl who has been treated since the age of one month for hip dysplasia.
Arthrography showed increased capacity of the capsule and wider range of movement of the femoral head. Both views are in neutral position; (a) with telescopy and (b) with traction. Note (i) 'pooling' of contrast; (ii) descent of head on traction (arrow points to medial aspect of femoral head) and (iii) labrum is not well outlined.

when the femur is rotated laterally. Manipulation of the hip during arthrography will show that in abduction the hip joint is in normal position and the labrum well demonstrated.

Partial displacement

Partial displacement is another term for subluxation; there is no soft tissue interposition between the femoral head and the acetabulum.

Radiography may be within normal limits, especially if the radiograph is taken with the hip in neutral or 'frog' position; or there might be some 'stand off' of the ossific centre; but such an abnormal position becomes more obvious on stress examination.

Arthrography (Fig. 32) will show that the capacity of the joint capsule is often increased; that there is an increased range of movements of the femoral head, and pooling of the dye is usually seen. Such subluxation is usually best demonstrated by

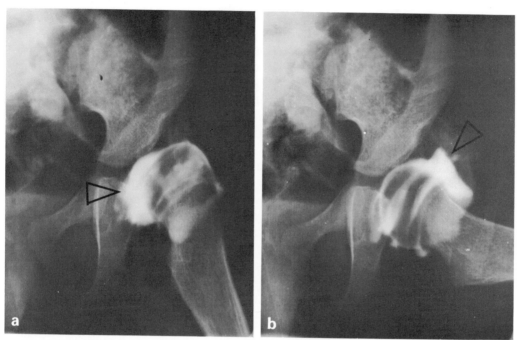

Figure 32 Partial displacement. Left hip arthrogram in a five-month-old girl, showing partial displacement of the femoral head with marked 'pooling' of the contrast material medially; this is best demonstrated on telescopy (a – see arrow). Also note that in this position the limbus is squashed and the 'rose-thorn' projection not present. However, in the 'frog' position (b) the femoral head is well seated and the pool of contrast is now squeezed in the supero-lateral aspect of the capsule and the limbus has reappeared.

pushing the leg upwards when it displaces and squashes the limbus against the acetabulum; then as the femur is abducted the femoral head slips back in the socket and the limbus reappears.

Complete displacement

This forms the most severe group of these cases. An example shows true displacement between the acetabulum and the femoral head. As the head slips out of the acetabulum it tends to cause the labrum to become inverted and to be pushed to lie posteriorly between the head and the posterior wall of the acetabulum.

Properly performed plain radiographs usually show abnormal hips but the degree of displacement depends on the type of the dislocation – 'tight' or 'loose'.

(a) Tight dislocation – On the plain radiograph the displacement of the femoral head is not marked and it may be mistaken for partial displacement.

Arthrography will show the soft tissue interposition; this is due to the inturned limbus, shown as a filling defect in the posterior aspect of the joint, as it slips behind the femoral head. As a result the 'landmarks' are lost and if proper views are taken the filling defect obstructing the head can be well shown (Fig. 33).

(b) 'Loose' dislocation – Radiograph of the hip shows obvious displacement which is much more marked than that of the 'tight' variety. The 'stand-off' now becomes true lateral displacement which is shown by the increased gap between the acetabulum and the head of the femur. The cephalad displacement or high position of the femur is especially evident in this 'loose' variety. These types of displacement may be associated with hypoplasia of the capital ossific centre.

Arthrography will show that the capacity of the capsule and the pooling of contrast medium tend to increase with the extent of the displacement. It will also show that the inturned limbus has now slipped behind the femoral head and is responsible for this displacement (Fig. 34). In more advanced cases, arthrography will show a constriction of the capsule between the femoral head and the acetabulum (Fig. 35); this is due to the capsular infold which is inverted with the limbus giving the so-called 'hour-glass constriction'. The high position of the head of the femur is usually associated with such an hour-glass constriction of the joint capsule (Fig. 36).

8.2.2   *Dislocation associated with persistent generalized joint hypermobility*

Laxity of the capsule is well demonstrated by arthrography; it shows increased capacity. Whereas in the average normal case, on injecting a few ml of saline into the joint, the pressure is increased so much that on releasing the plunger of the syringe, the saline will flow back in the syringe; in cases of capsular laxity such backflow is not detected and the capsular capacity is much increased. In hypermobile but non-dislocatable joints the femoral head is well-developed and spherical and so also the acetabulum. The articular surfaces are smooth. Likewise the labrum is present and gives adequate coverage to the head. The femoral head, although it shows some increase in the range of 'play' within the acetabulum, is well retained within the acetabulum throughout the range of movement.

If a hypermobile hip joint is unstable it can be dislocated; a feature of such a hypermobile and dislocating joint is that it can be completely reduced. Such a loose and wide displacement on dislocation is made possible by the elongation and bagginess of the capsule and ligaments; the labrum retains its normal position and is

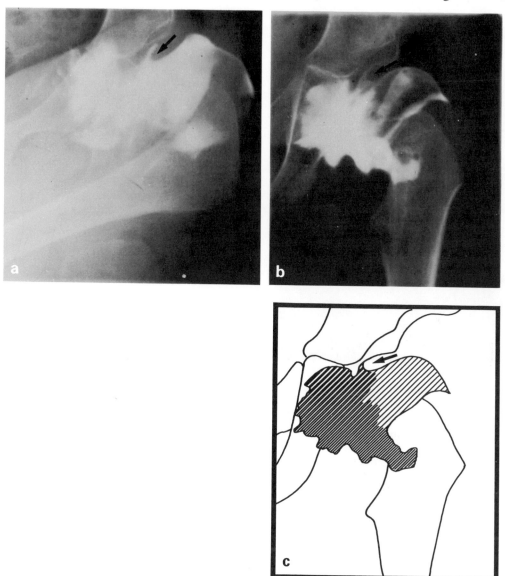

Figure 33   Complete dislocation of the tight variety. Left hip arthrogram in a
ten-month-old girl showing an inturned limbus (see arrow) obstructing the head of the
femur. (a) Adduction and flexion; (b) neutral position and (c) line drawing of b.

well-maintained, and it does not become inverted on dislocation, nor is there any
interposition of soft tissue between the head and acetabulum.

When this joint laxity is severe, especially if there is a history of repeated
dislocation, articular erosions and degeneration can take place. If such degenerative

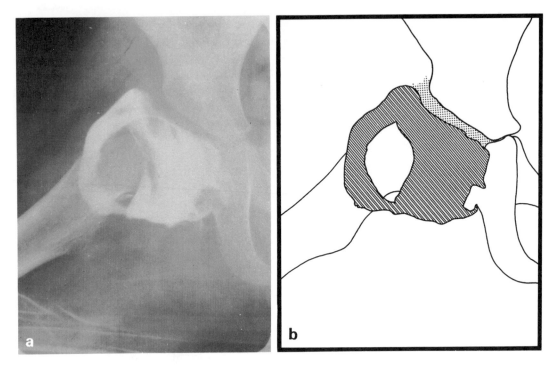

Figure 34   Complete displacement of the 'loose' variety. (a) Right hip arthrogram on a fourteen-month-old child showing a capsule of increased capacity with loss of 'parallelism' due to fibro-cartilaginous interposition between the femoral head and the acetabulum. (b) Line drawing explaining the various structures.

erosive changes have occurred, arthrography will show uneven articular surface of the femoral head.

The arthrographic appearances of such persistent joint laxity involving the hips was illustrated in a case which was considered to be an example of arthrochalasis multiplex congenita associated with Down's syndrome (Owen, Elson and Grech, 1973).

Case Report (Case 6)

A.F., a boy, was born at home but admitted to the Premature Baby Unit because of low birthweight (2·1 kg) where features of Down's syndrome were noted and the diagnosis was confirmed by chromosome analysis. He was discharged home when 12 days old and was followed up as an out-patient.

Follow-up showed the usual course of Down's syndrome but otherwise was uneventful for the first six years of life. It was then reported by the mother that she found him in the morning lying with his left hip acutely flexed and with a swelling apparent in his groin. She also reported that she was able to correct the position of the hip by manipulating the thigh and that such correction was accompanied by a loud and

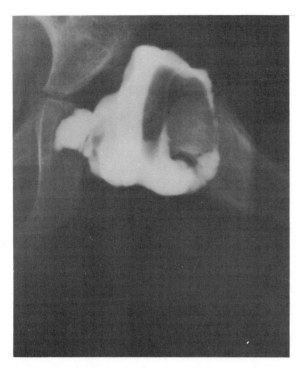

Figure 35   Complete displacement of the right hip of the 'loose' variety in a girl of just over a year of age showing an early hour-glass constriction of the capsule.

palpable click.

Preliminary examination revealed muscle hypotonia and also definite hyper-extensibility of other joints. His gait was plodding in character, but he could run and stand on either leg unaided.

He was admitted to hospital for full evaluation including bilateral arthrography. This showed severe laxity of both capsules with a range of movement above normal. There was no other anatomical abnormality. It was found that antero-inferior dislocation in both hips could easily be produced by thrusting downwards the femora with the knees and hips flexed. Subsequent reduction was easy and complete by relaxing the downward push and extending the hips and knees into neutral position (Fig. 37).

The articular surfaces of both femoral heads can be seen to be smooth and even and there is no evidence of any degenerative erosive changes.

No active measures were advised and the boy continues to be observed as an out-patient. He is now 11 years old. The episodes of dislocations gradually decreased in frequency – over the last three years (1972–75) there was an average of one episode of hip dislocation per year. On each occasion reduction of the dislocation was easy. All these episodes occurred at night while asleep, except for the latest incident (3.7.75) when he fell at school while playing. The cause of this gradual improvement is not apparent, but obviously the hip joints are becoming more stable and one cannot help speculating that the retaining powers of the ligaments and capsule are becoming more competent.

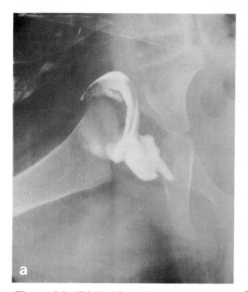

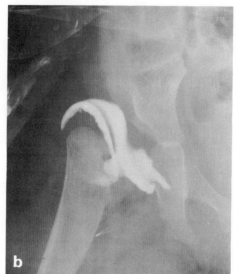

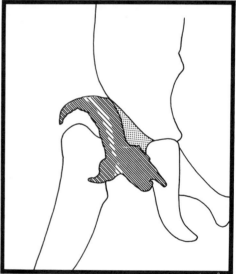

Figure 36   Right hip arthrogram on a girl of nine months showing complete dislocation with lateral and cephalad displacement of the head with a well-established hour-glass constriction. There was so much displacement of the femoral head that it was considered easier to enter the joint space medially to the femoral head. Note the obliteration of the 'landmarks' of the arthrogram, the high position of the zona orbicularis and the early formation of a 'false acetabulum'. The schematic representation explains the arthrographic findings.

Figure 37   Case 6   (a) Bilateral hip arthrograms in neutral position showing capsular laxity but otherwise both femoral heads appear to be normally seated. (b) Both hips are flexed and downward thrust is applied on the flexed knees producing complete dislocation of both hip joints. Note the hyper-elasticity of the joint capsules. Reduction was easy and complete. (c) Radiograph of the pelvis after the latest incident (3.7.75) following a fall while playing at school showing posterior dislocation of the left hip which was again reduced easily.

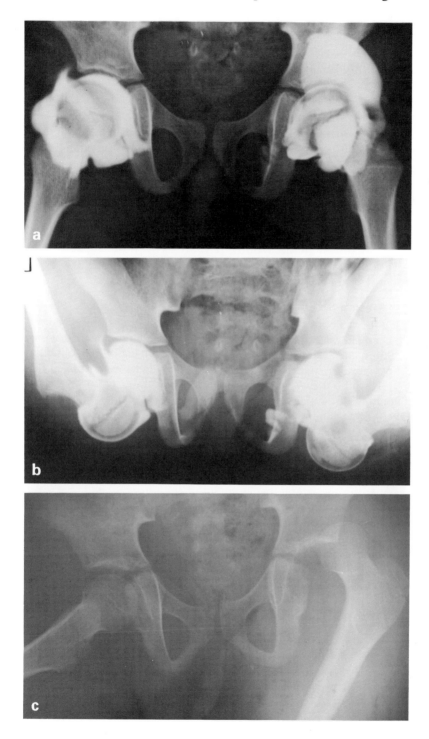

8.2.3   *Paralytic dislocations of the hip*

Subluxation or dislocation can take place in the hip of infants suffering from spastic hemiplegia, poliomyelitis or spina bifida cystica.

There are characteristic bone changes as a result of the cessation of function of the limb due to the paralysis, which include decalcification with thinning of the cortex of the bones of the affected limb. Also the angle of the femoral neck with the shaft is increased. The normal angle at birth is about 160 degrees and it decreases to about 125 degrees in adult life. In the paralysed leg, this angle may increase to approach 180 degrees. Therefore the factors responsible for such hip dislocations are the muscle imbalance and the resulting developmental defects in the joint, i.e. coxa valga and shallow acetabulum.

The arthrographic findings are often the same as those seen in hip dysplasia. The femoral head is often spherical but the coxa valga deformity is characteristic of the paralytic subluxation in the young patient. The joint capsule is lax with a large capacity. The acetabular labrum is often elevated but it can also be inverted making the dislocation non-reducible.

Case Report (Case 7)
T.C., a girl, was born on 31.7.67 with a lumbosacral myelomeningocoele, which was closed when she was a few hours of age. When the wound healed she was discharged home. Later, the mother noted that she kicked both legs. On a follow-up examination both hips showed passive abduction of 90° and there was no fixed flexion. Internal rotation was nil but there was 90° of external rotation.

Radiograph of the pelvis of 23.1.68 showed marked lateral displacement of the left capital epiphysis. She was put on a Von Rosen splint but repeated X-rays showed that the left hip continued to stand off. By this time the mother also noted that the right leg was more active than the left. Besides this lateral displacement, there was also some telescoping of the left hip on examination. Some stabilizing procedure was considered necessary and as a result left psoas transplant operation was carried out. There was no improvement in the lateral displacement of the femoral head (Fig. 38, a).

Left hip arthrogram was performed on 17.9.71 on account of this persistent hip displacement. This confirmed the hip dislocation and that reduction was prevented by an inturned limbus (Fig. 38, b, c and d).

On open reduction, the limbus was found to be inturned; consequently limbectomy was carried out followed by pelvic osteotomy.

This gave satisfactory stabilization; the child is now taking weight on both legs with minimal support. As her quadriceps power is considered satisfactory, she will be started with below knee calipers.

We have come across a case of arthrogryposis multiplex congenita, which is included here for convenience sake. Arthrogryposis is a neuromuscular syndrome characterized by congenital multiple joint contractures associated with muscular deficiencies, and a progressive deposition of fat within the muscle. Hip dislocation is not uncommon; it occurred in 17 out of 37 cases (approximately 46%) where

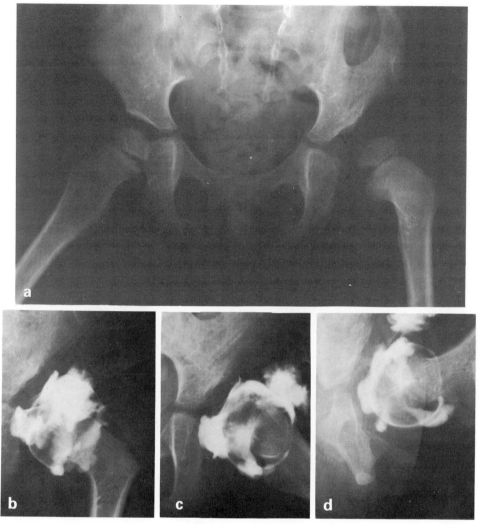

Figure 38   Case 7   (a) Antero-posterior view of the pelvis of a girl of four years with an extensive lumbo-sacral myelomeningocoele which was closed soon after birth, showing that left psoas transplant operation was carried out. In spite of this stabilizing procedure, there is lateral displacement of the left femoral head. (b, c and d) Arthrographic study in neutral, abduction and rotation and in frog positions respectively, confirming the hip dislocation due to an inturned limbus which is obstructing the head.

radiographs were available in a series of 41 cases reviewed by Poznanski and La Rowe (1970). They found that when dislocation was frank, reduction was difficult or impossible.

Case Report (Case 8)

L.W. (Fig. 39), a boy, was born on 24.5.73 by caesarean section for prolapsed cord with breech presentation. He presented a typical clinical picture of arthrogryposis including bilateral club feet, dislocated right hip (Fig. 40, a) and flexion contractures of wrists and knees. Closed reduction of the dislocated right hip was attempted but failed. He was referred for bilateral hip arthrograms as open reduction was contemplated.

Bilateral arthrography (Fig. 40, b and c) was carried out on 25.10.74 and this showed

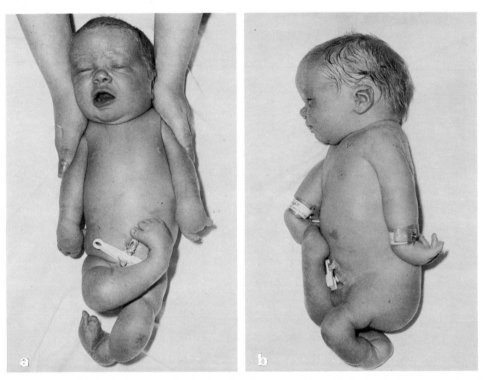

Figure 39   Clinical photograph of Case 8, taken soon after birth. Note the generalized flexion contractures and deformities of most of the limb joints including bilateral clubfeet. (By courtesy of Mr J. F. Patrick, F.R.C.S.)

Figure 40   Case 8   (a) Antero-posterior view of the pelvis, showing hypoplasia of the right capital ossific centre with cephalad and lateral displacement; the left hip joint appears normal. (b and c) Bilateral hip arthrograms showing frank dislocation of the right hip which could not be reduced on screening; the capsule was rather inelastic and very little contrast could be injected; although the articular cartilage of the right femoral head appeared smooth, the head was smaller than the left one. The left arthrogram is normal. At operation the dislocation of the right hip was confirmed to be irreducible; the acetabulum was found to be full of fat which had to be cleared before the femoral head could be properly seated.

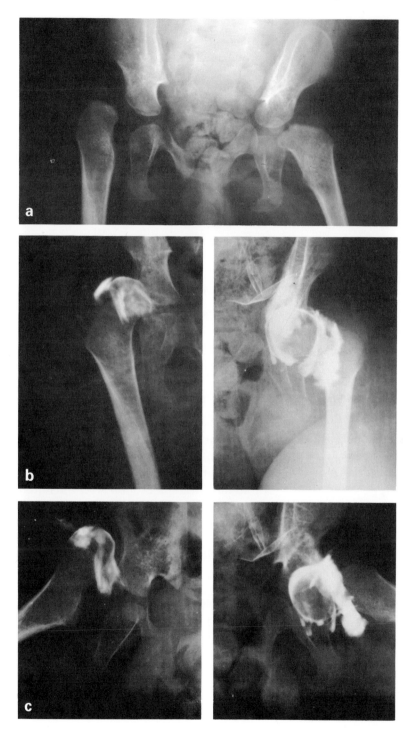

a normal and stable left hip but a dislocated and irreducible right hip. The dislocated right femoral head was smaller than the left, but the articular cartilage appeared normal. The capsule, however, was small, fibrotic and inelastic.

When open reduction was carried out, the muscle tissue around the hip joint was found to be deficient. In many places there was simply fat and fibrotic tissue in place of the majority of the muscle. When the hip was opened, the acetabulum was found to be full of fat and the hip was irreducible until the acetabulum was cleared of this excess fat.

8.2.4   *Arthrographic appearances in Perthes disease*

Both radiographic and arthrographic findings in Perthes' (Legg–Calvé–Perthes) disease depend on the stage of the illness. Detailed description of the radiographic changes is outside the scope of this book, but these are summarized together with the arthrographic appearances in Table 6.

Perthes disease is usually described in three stages, each stage lasting approximately nine months or more; the duration of each stage is variable, being quite short in some younger children. Such subdivision of the disease has appeared under different headings and I prefer to follow the Ozonoff terminology (1975):

(a) Early stage   – small epiphysis, sclerosis, fissuring and apparent medial displacement;

(b) Intermediate
stage   – resorption, followed by simultaneous resorption and re-ossification;

(c) Late stage   – complete reossification and remodelling of the trabecular pattern.

Using such terminology one avoids terms in common use like 'fragmentation' and 'reparative' stages, which can be misleading; 'fragmentation' gives the impression that the epiphysis itself is crumbling, whereas the cartilage is quite intact; also one can object to the term 'reparative' on the ground that all changes seen after the initial insult, whatever they are, are in actual fact, forms of repair.

Early stage
Waldenström (1938) described three early radiological features: an increase in the medial joint space; flattening of the femoral head and increased density of the capital epiphysis. According to Kemp and Boldero (1966) the lateral displacement of the femoral head, or better the increased width of the joint space, is the earliest radiological sign; they maintain that the resulting hypertrophic changes in the synovia ligamentum teres and capsule, demonstrable by the filling defects on the medial side of the joint space, are responsible for such displacement of the head of the femur; but others including Ozonoff (1975) have not found any significant evidence of such intra-articular filling defects in Perthes disease, and consequently the cause of such displacement is not yet fully known.

Before flattening of the ossific centre becomes apparent, one can often see a marginal fracture in the supero-lateral quadrant of the ossific centre, best demonstrated in the lateral position (Fig. 41, a).

Table 6   Classification and summary of the radiographic and arthrographic findings in Perthes' disease

Stage	Radiography	Arthrography
Early	Marginal fracture in ossific centre (Fig. 41). Increase in joint space. Flattening of the epiphysis. Metaphyseal rarefaction.	Femoral head may show normal surface and normal sphericity; minimal flattening of the cartilage may be seen but the arthrogram is close to normal (Fig. 41, b and d).
Intermediate	Ossific centre appears mottled and fragmented.	Often there is some increase in the capsular capacity but the cartilaginous contour of the femoral head is often well-maintained, despite the appearances of the ossific centre (Fig. 42). Sometimes the femoral head becomes larger, most probably depending on the extent of fragmentation. Such enlargement of the femoral head is reflected in changes in the thickness of the acetabular articular cartilage which may lead to incongruity (Fig. 45)
Late	Regeneration of epiphyseal outline and gradual disappearance of the dense foci in the capital epiphysis.	The caput index is the least at this stage accounting for the femoral head deformity which may be significant; there may be incongruity and lack of proper coverage. Arthrography will show the regeneration of the articular cartilage to normal thickness (Figs 43 and 44).

The arthrogram shows the femoral head to have a smooth surface and usually, in spite of the obvious changes in the ossific centre, as seen on the plain radiograph, the arthrographic findings are close to normal (Fig. 41, b); but most cases show minimal flattening of the cartilage directly over the fissure (Fig. 41, c, d). According to Ozonoff (1975) only one or two cases of his series have been truly normal and flattening is a characteristic finding. On the whole, arthrography has little to contribute at this stage.

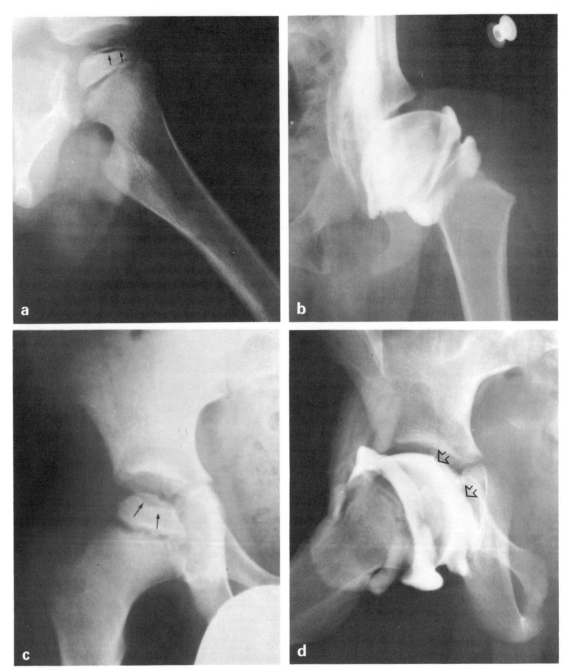

Figure 41   (a) Early Perthes disease of the left hip in a three-year-old boy with a short history of two weeks of regional pain and limp. A frog view of the left hip showing minimal flattening of the ossific centre and a sub-cortical fracture under the superolateral aspect of the ossific centre (see arrows). This is best demonstrated in a lateral position. (b) Left hip arthrogram showing a normal joint. (c) Radiograph of the right hip of a six-year-old boy who has been complaining of pain and a limp for three months, showing fissuring of the capital epiphysis (small arrows). (d) Right hip arthrogram on this boy shows flattening directly over the fissure (open arrows). This finding is characteristic of early Perthes disease.

Intermediate stage
On the plain radiograph, the capital ossific centre appears mottled and fragmented (Fig. 42, a).

The arthrographic findings depend on the extent of the fragmentation – these can vary from almost normal appearance to those seen in the late stage, depending on

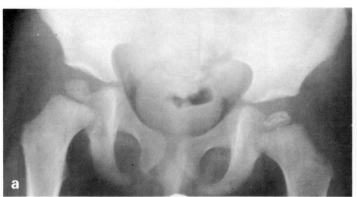
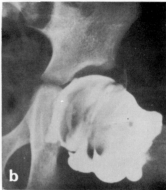

Figure 42   Well-established radiological changes of Perthes disease in a two-year-old boy. (a) A standard antero-posterior view of both hips showing flattening and fissuring of the left capital epiphysis with increased width of the joint space. (b) Left hip arthrogram performed a few weeks later, showing the cartilaginous contour of the femoral head is well-maintained. On arthrocentesis, some fluid was aspirated and compared with Fig. 41a there is increase in the capacity of the capsule. It is presumed that this synovial effusion could account partly for the lateral displacement of the femoral head.

how long the patient has been walking on this softened head. In the later phase of fragmentation of the epiphysis, there may be some enlargement of the femoral head which retains a regular articular surface. Such femoral head enlargement is accompanied by some gradual alteration of the acetabular articular cartilage which becomes narrower laterally and slightly widened medially; this is the beginning of incongruity of the joint if this is going to occur. The most extreme abnormality would be a lateral extension of cartilage out under the labrum to form a 'bump'. Most cases will show a more ovoid than round head, with few of them as extreme as the ones showing the 'bump' deformity.

The arthrogram shows the changes in the contour of the articular cartilage which cannot be demonstrated by plain radiography. It will confirm that the cartilage is usually quite smooth and in no way conforms to the apparent fragmentation of the osseous epiphysis.

Late stage
This consists in the regeneration and reossification of the epiphysis. As a result there is gradual disappearance of the dense epiphyseal foci until finally there is complete

reformation of the epiphysis (Fig. 43, a). The findings in this last phase are merely a confirmation of those indicated in the second phase. As the cartilaginous model is set during the intermediate phase, there really can be little change in the late phase (Fig. 44).

The arthrographic findings depend on the extent of congruity. When reconstitution occurs with minimum deformity, the arthrographic findings can be within normal limits (Fig. 43, b). On the other hand there may be marked enlargement of the femoral head and obvious incongruity between it and the acetabular socket. In the neutral and adduction positions, such incongruity is not too obvious, unless there is a large lateral bump; in abduction, however, marked incongruity can be shown which can be quite striking. When there is extreme incongruity, the head fulcrums around the lateral deformity opening up the joint space medially (Fig. 45). There may also be compression of the lumbus with lack of coverage of the femoral head; this compression will lead to narrowing of the lateral aspect of the joint space and it may cause depressions into the femoral head, usually laterally.

Consequently arthrography can give some useful information in the fragmentation-regeneration phase of the disease. It shows the resulting shape of the femoral head, and whether the lateral part of the head is protruding beyond the acetabular labrum or not; it will also exclude any indentation by the labrum on the surface of the femoral head.

Jonsäter (1953) and later Katz (1968) analysed their arthrograms by calculating the ratio of the femoral head to half the transverse diameter, as an expression of the sphericity of the femoral head. There was good correlation between the two studies. It was found that the values for the caput index was nearest to normal in the initial stage, somewhat less in the intermediate stage and least in the late stage.

The use of arthrography in Perthes disease has been aptly summarized by Katz (1968):

'Arthrography cannot be considered part of the routine workup and follow-up in patients with coxa plana. It adds little information in the earliest (initial) or latest (definitive) stages; in the one, the contrast substance shows a virtually normal articular outline, and in the other there is sufficient healing of the ossific nucleus to allow its border to parallel that of the articular cartilage.

In the fragmentation and reparative stages, the following variation in the arthrogram may yield important information:

(a) Confirmation of normal or near normal reconstitution of the femoral capital epiphysis with good congruity with the acetabulum;
(b) Enlargement of the femoral head, not adequately demonstrated by plain films;
(c) Poor congruity between femoral head and socket with altered acetabular cartilage thickness;

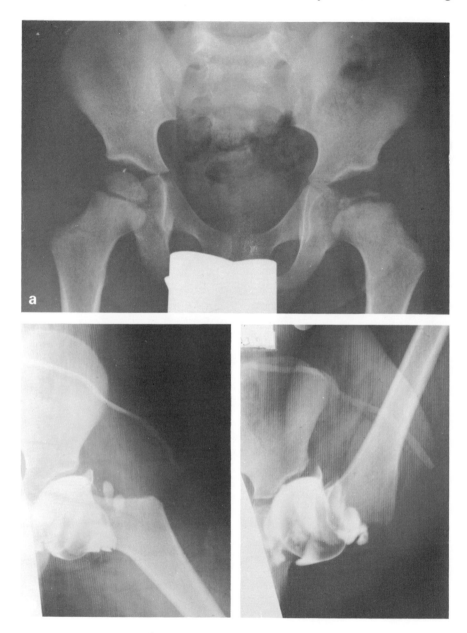

Figure 43   Regeneration stage of Perthes disease in the left hip of a boy of four years.
(a) Standard antero-posterior view of the pelvis showing the ossific centre of the left hip
is in late stage of fibrous replacement. (b) Two arthrographic views showing the joint
space is well-outlined and the cartilaginous head has a smooth outline and the caput
index is normal. There is no evidence of incongruity.

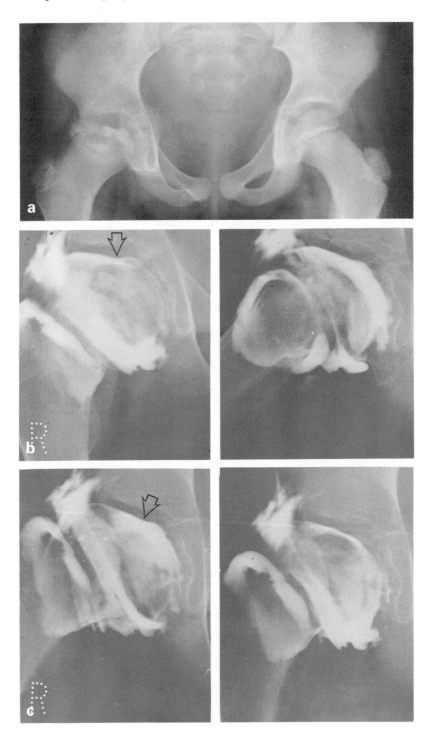

Figure 44   Perthes disease in the late reparative stage in a ten-year-old girl. Top is an antero-posterior view of the pelvis. The right femoral head epiphysis is flattened and irregularly sclerotic. The width of the joint space is increased medially and radiolucent area is seen extending into the metaphysis of the femoral neck which is widened. Arthrographic views of the right hip showing some flattening of the femoral head; this flattening is particularly noticeable on the superolateral aspect (see arrow); in spite of this flattening the head is well contained.

Figure 45   Arthrograms on three children with Perthes disease to show the varying degree of incongruity between the femoral head and acetabulum. (a) Right hip arthrogram on a six-and-a-half-year-old-boy showing minimal incongruity. (b) Shows moderate degree in a nine-year-old boy. (c) Right hip arthrogram on a six-year-old girl demonstrating marked incongruity. Note that incongruity of the joint is best shown in abduction. When such incongruity is extreme, the head fulcrums around the lateral deformity or 'bump', opening up the joint. This is best demonstrated in c. (Courtesy of Dr M. B. Ozonoff, Connecticut).

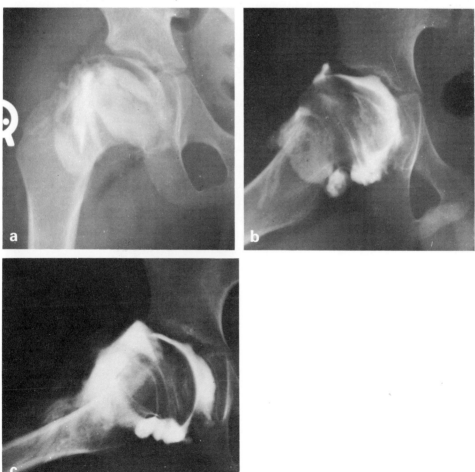

(d) Poor congruity between the femoral head and the socket with depressions in the head by the fibro-cartilaginous labrum or acetabular margin or both'.

8.2.5   *Following total hip replacement*

After total hip replacement, a fibrous pseudocapsule forms around the free part of the prosthesis that is not covered by the acrylic cement or bone. This pseudocapsule extends from the rim of the prosthetic socket to the base of the femoral neck, forming an irregular space. Such an '*intracapsular*' space should not communicate with the surrounding soft tissue structures or bursae.

The technique used in these cases is somewhat modified from that described (Chapter 3). It is carried out under local anaesthesia with strict aseptic precautions in the X-ray department. A 20–22 gauge disposable spinal needle with a short bevel is introduced, and the needle is advanced under image-intensifying fluoroscopy to hit the neck of the prosthesis. Since the prosthesis is metallic, it may superimpose the image of the needle and thus interfere with its visualization. Consequently it has been found helpful to introduce the needle more laterally than usual.

As the needle hits the prosthesis, the metal to metal contact is obvious. Aspiration is then attempted and what is aspirated should be sent to a laboratory for examination. When the content of this pseudocapsule is aspirated, the syringe is replaced by another containing water-soluble contrast material, and this is gradually injected under direct vision until this intracapsular space is filled. Usually about 7 ml of contrast are needed, but this depends on the size of this 'pseudo-capsule' and also on the presence or otherwise of any abnormal communications.

When this para-cervical space is adequately filled with contrast medium, the needle is withdrawn and the hip joint is actively or passively exercised. Overcouch radiographs of the hips are then taken ensuring that the whole area of the prosthesis, including the acetabular and the whole shaft, are included in the radiograph. Antero-posterior view is usually enough, but in some cases oblique or lateral projections may be needed. This examination must also include an antero-posterior radiograph while traction is applied to the leg (Fig. 46).

When the prosthesis is well-seated and firm, any contrast medium injected in the pseudo-capsular space will fill an irregular space around the neck of the prosthesis, but it will not run down along the shaft of the prosthesis into the acrylic – bone interface. Often a fine radiolucent line along the acrylic layer is seen; unless such a line widens on serial examination or with traction or unless the contrast medium seeps into it, it is of no pathological significance.

On the other hand if loosening of the prosthesis is present, there is bone resorption along the shaft and a gap is created along the acrylic – bone interface. On injecting the contrast in the pseudocapsular space, it will run into the acrylic – bone gap. Applying traction to the leg will enhance such seepage of contrast along the shaft of the prosthesis (Fig. 46).

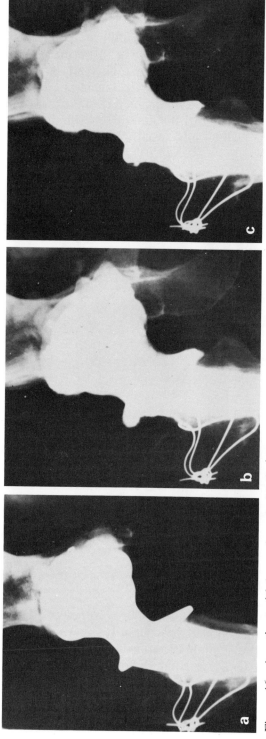

Figure 46 A patient with loosening of a McKee Farrar prosthesis on the acetabular side. (a) Plain antero-posterior view showing some bone resorption on the acetabular side suggestive of a loose cup. (b) Arthrogram – antero-posterior view without traction. (c) Arthrogram – antero-posterior view with traction. The contrast medium is now more clearly seen, confirming the loosening of the cup. (Courtesy of Dr. E. A. Salvati, New York).

Interpretation of the arthrogram can be complicated in those cases where the acrylic cement used is radiopaque. In these cases where there is loosening of the prosthesis, there is superimposition of two opaque shadows – the cement and the seeping contrast medium, and it may become difficult to tell them apart.

This can be solved by means of the *subtraction* technique (Salvati *et al.*, 1974) (Fig. 47). Subtraction was first introduced by Ziedses des Plantes in 1934. A positive transparency is made of a radiograph, usually the control radiograph. A second

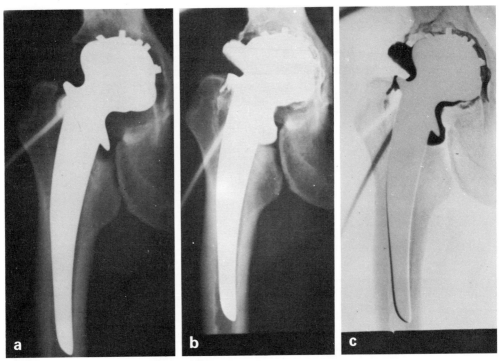

Figure 47   Subtraction technique in arthrography following total hip replacement. (a) Control. (b) Arthrogram identical position to the control. (c) Subtraction – cancelling out shadows identical to both radiographs (a and b). As a result the contrast medium present in b stands out more clearly (Courtesy of Dr D. J. Stoker, London).

radiograph, identical to the first except for the presence of additional data, in this case the contrast material introduced, is accurately superimposed upon this positive transparency. Cancelling out details common to both radiographs, will make the contrast medium present in the second radiograph stand out more clearly and consequently the diagnosis is more obvious. Subtraction technique has been widely used in vascular and neuro-radiology and it has now been shown to be also useful in arthrography for showing loosening of hip prosthesis fixed with radio-opaque cement. This technique subtracts the preliminary from the post-injection radio-graphy thus showing the injected contrast agent. For this technique to be useful, the pre-injection and post-injection radiographs must be exactly comparable and therefore can be perfectly superimposed. Consequently the patient must not move or be moved during the examination.

8.2.6   *Following trauma and degenerative changes/localization of loose bodies*

The trauma associated with the production and reduction of hip joint dislocations is often responsible for soft tissue damage, involving not only the vascularity of the proximal end of the femur, but also the capsule. Such dislocations can be accompanied by other complicating associated lesions including fractures of acetabulum, femoral head or trochanter. As a rule the hip joint in the child is easier to dislocate than in an adult, but the injury is complicated by a much lower incidence of associated lesions (Donaldson *et al.*, 1968). The incidence of degenerative changes resulting from such dislocation appears to be less than in the adult. On the whole such traumatic hip dislocations are less troublesome in the child than in the adult. Maybe that is why all the cases that fall under this category which were referred for arthrography were adults.

The following is an example of this type of case.

Case Report (Case 9)
A thirty-one-year old man sustained a fracture dislocation of the left hip; the dislocation was associated with a marginal fracture of the roof of the acetabulum. After reduction the hip remained unstable, so open reduction was carried out, together with screwing the acetabular fragment back in position; at operation this piece of acetabulum was found to amount to about a third of the articular surface.

Hip arthrogram (Fig. 48) was requested six weeks after the operation to confirm the joint stability and the integrity of the joint capsule. Although there was some limitation of movements, the head was well-seated and stable and there was no extravasation of contrast medium suggesting capsular tears.

Arthrography has also been utilized in a few cases to investigate the possibility of loose bodies in the hip-joint, where plain radiography was inconclusive.

Case Report (Case 10)
A girl of fifteen years with a history of old congenital hip dysplasia was referred for arthrography on account of limitation of movements in the left hip, together with pain and a history suggestive of 'locking'. A plain radiograph done in another hospital referred to the changes in the acetabulum and the femoral head due to the old disease, and also commented on the 'loose' fragment off the superolateral margin of the acetabulum. The referring orthopaedic surgeon was not convinced that this ossicle was extra capsular and most probably represented the os acetabulare. As the girl continued to complain, arthrography (Fig. 49) was carried out under local anaesthesia. This

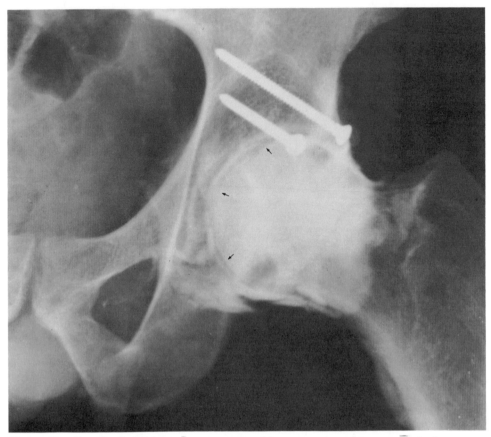

Figure 48   Case 9   Left hip arthrogram on a thirty-one-year old man who had sustained a fracture dislocation of the left hip six weeks previously. Arthrogram shows the hip to be in good position and stable; there was no radiological evidence of any capsular tears. Note the articular surface of the femoral head (arrows) to be smooth and even, appearing as a regular translucent linear shadow between the ossified head and the contrast layer in the joint space.

confirmed that the bone fragment was outside the joint and that it retained the same relationship to the acetabular margin suggesting that it was not 'loose'. It also showed that although the acetabulum was rather shallow and the head somewhat flattened, the latter was well seated and the joint was stable.

8.2.7   *Pyogenic arthritis of the hip*
Following suppurative infections of the hip joint, marked destruction of most of the components of the joint can follow which may lead to dislocation.

The joint capsule becomes thickened and its elasticity disappears; as a result the

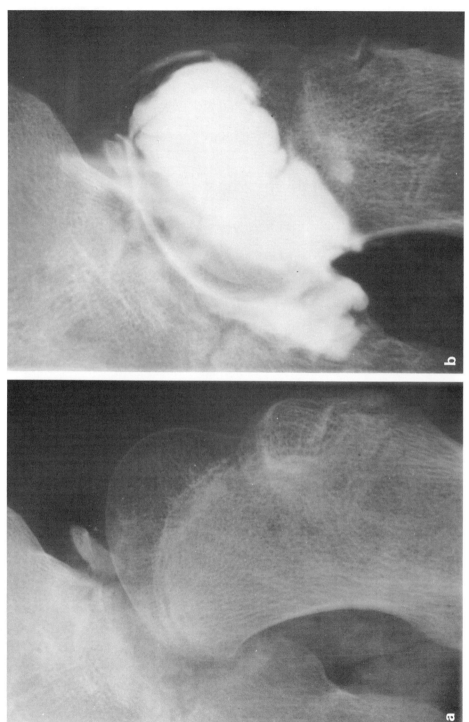

Figure 49 *Case 10* (a) Radiograph of the left hip showing very shallow acetabulum (the acetabular index was found to be only 0·3) with some coxa valga deformity and the femoral head in a position of subluxation. There is obvious disparity between the size of the head and the socket. There was some doubt clinically about the bone fragment opposite the lateral margin of the acetabulum. (b) Arthrogram confirms that this bone fragment is extra-capsular and that it is not 'loose'; it also showed that in spite of the disparity in size, the head was stable.

capacity of the joint capsule is greatly reduced. The arthrographic outline is usually irregular due to the fibrosis and adhesions that result.

The femoral head is often eroded and deformed and the smooth articular surface becomes irregular and pitted and consequently the femoral head is smaller than normal.

Filling defects can be seen in the cartilaginous part of the acetabulum. With dislocation the usual landmarks of the arthrogram cannot be identified. Also such a small deformed dislocated femoral head is often fixed and cannot be reduced.

The following is an example of this type of case.

Case Report (Case 11)
P.A., a boy, was born after medical induction three weeks before being due. The birth weight was 7 lb 14 oz. He was well for the first two weeks of life, when he started to vomit, and on examination he was found to have a tender swelling over his left hip. He was re-admitted; aspiration of the left hip grew B. haemolytic streptococcus. The following investigations were carried out:

Hb.	11·9 G/100 ml
W.B.C.	16 000 mm^{-3}; lymphocytes – 13 280 mm^{-3}
Urine	No cells; no growth
Urea	26 mg percent
Electrolytes	normal
Blood cultures	no growth

X-ray left hip showed slight new bone formation around left femoral neck, but was otherwise normal.

The child was put on penicillin and Cloxacillin, the swelling subsided and he was discharged.

Soon after the parents left England as the father worked abroad. Whilst abroad the child was examined by a paediatrician who thought that the child had a 'dislocated left hip'. The mother brought the child back to England as she was anxious for a second opinion.

By the time he was admitted to this hospital the child was five months old. There was limitation of movement in the left hip, but otherwise he was symptom-free. Plain radiograph of the hips showed a normal right hip, but the capital epiphysis on the left side was smaller than the right, and the head appeared to be dislocated. Also there was some new bone formation near the lesser trochanteric area.

Arthrography of the left hip was requested. This (Fig. 50) showed an irregular, non-elastic capsule around a small dislocated head. The head was also shown to be uneven, and it was not possible to reduce it; filling defects could be seen in the acetabulum which were preventing the proper seating of the head.

These arthrographic findings are typical of dislocation following septic arthritis. This investigation not only confirmed the complete dislocation of the hip joint, but also that the femoral head, although small and somewhat irregular, was worth saving by exploration of the joint and open reduction to be followed by Salter's osteotomy for stabilization.

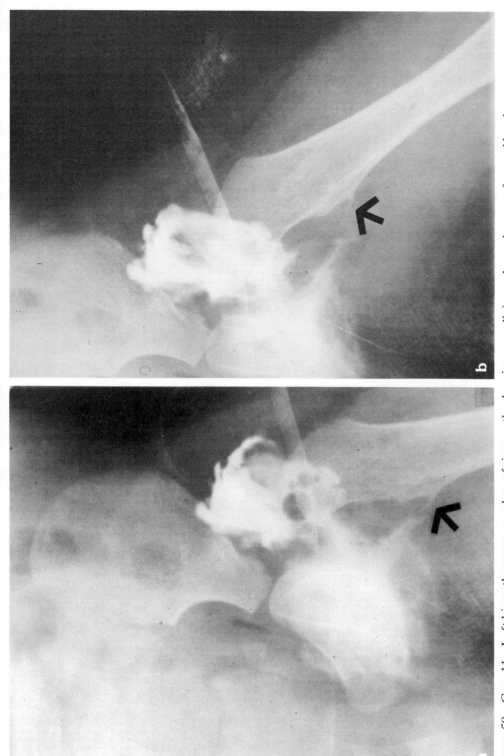

Figure 50   Case 11   Left hip arthrogram on a boy of six months showing a small, irregular and uneven femoral head which is dislocated and irreducible. The capsule was of small capacity and did not feel elastic on injection, suggesting that the capsule was fibrotic; also filling defects can be seen between the acetabulum and the head. At open reduction a fibrosed hip joint with a dislocated femoral head following septic arthritis was confirmed. (a) neutral position; (b) abduction with attempted internal rotation. Arrow points to the periosteal reaction along lesser trochanter.

References

Astley, R. (1967), Arthrography in congenital dislocation of the hip. *Clinical Radiology*, **18**, 253–260.

Donaldson, W. F., Rodriquez, Skovron, M. and Gartland, J. J. (1968), Traumatic dislocation of the hip in children. Final Report by the Scientific Research Committee of the Pennsylvania Orthopaedic Society. *Journal of Bone and Joint Surgery*, **50–A**, 79–87.

Felländer, M., Gladnikoff, H. and Jacobson, E. (1970), Prevention of congenital dislocation of the hip joint in Sweden — efficiency of early diagnosis and treatment. *Acta Orthopaedica Scandinavica*, Supplement 130, Munksgaard, Copenhagen.

Grech, P. (1972), Arthrography in hip dysplasia in infants. *Radiography*, **38**, 172–179.

Jonsäter, A. (1953) Coxaplana, a histopathologic and arthrographic study. *Acta Orthropaedica Scandinavica*. Supplement 12. Munksgaard, Copenhagen.

Katz, J. F. (1968), Arthrography in Legg–Calvé–Perthes disease. *Journal of Bone and Joint Surgery*, **50–A**, 467–472.

Kemp, H. S. and Boldero, J. L. (1966), Radiological changes in Perthes disease. *British Journal of Radiology*, **39**, 744–760.

Mitchell, G. P. (1963) Arthrography in congenital displacement of the hip. *Journal of Bone and Joint Surgery*, **45–B**, 88–95.

Owen, J. R., Elson, R. A. and Grech, P. (1973), Generalized hypermobility of joints: Arthrochalasis multiplex congenita. *Archives of disease in Childhood*, **48**, 487–489.

Ozonoff, M. B. (1975), Personal communication.

Poznanski, A. K. and La Rowe, P. C. (1970), Radiographic manifestation of the arthrogryposis syndrome. *Radiology*, **95**, 353–358.

Salvati, E. A., Ghelman, B., McLaren, T. and Wilson, P. H. (1974), Subtraction technique in arthrography for loosening of total hip replacement fixed with radiopaque cement. *Clinical Orthopaedics and Related Research*, **101**, 105–109.

Severin, E. (1939), Arthrography in congenital dislocation of the hip. *Journal of Bone and Joint Surgery*, **21**, 304–313.

Waldenström, H. (1938), The first stages of Coxa Plana. *Journal of Bone and Joint Surgery*, **20**, 559–556.

Ziedes des Plantes, B. G. (1934), Planigraphie en subtractie Röntgenographische differentiatiemethodem. Thesis, Utrecht.

9 Radiation hazard

with special reference to arthrography in children

Radiologists are becoming increasingly conscious of the radiation hazards. The hazards of radiography must be set against the value and information gained from the examination; when radiological examination is decided upon, steps must be taken to cut down the radiation to a minimum. Such irradiation of the patient must be no greater than is necessary to produce a satisfactory result, especially in hip radiography where the area X-rayed is so close to the gonads and since most of these patients are young children.

In such investigations, the radiation dose rate to the patient is related to (a) fluoroscopy and (b) radiography. Obviously the longer one screens the patient and the more radiographs that are taken, the greater will be the dose. To cut down such irradiation the following principles must be followed:

(a) The examination should be performed only for proper indications. Unnecessary examinations are sometimes requested due to failure to ascertain whether there are records of previous radiology. It is considered that such investigations should be requested by senior clinicians and that these clinicians must satisfy themselves that such examinations are necessary. Case discussion between clinicians and radiologists should be encouraged before the examination is undertaken.

(b) The procedure should be done with proper skill and the technique adopted should be such as to give the best possible information. This will eliminate inconclusive results or the need for repeated examinations.

(c) Such investigations must be carried out in the Radiology department. These radiographic suites must be adequately equipped. Image-intensification should be used in order to reduce the dose received by the patient.

(d) Fluoroscopy must be curtailed and likewise the number of radiographs should not be excessive. A record of the factors used and the number of exposures, together with the fluoroscopy time, should be kept and recorded. From this, one can approximately estimate the radiation dose given.

(e) All possible steps for reducing the irradiation, especially coning down to the smallest possible field size, should be encouraged.

(f) Whenever practicable suitable gonad shielding should be used. However as most of these patients are children, and in hip dysplasia usually girls, covering the gonads is often difficult.

(g) Mechanical devices to ensure immobilization should be used on small children; such supports will not only make the examination easier and quicker, but they ensure that others are not exposed to radiation unnecessarily.

In such a diagnostic procedure, if these protective measures are employed, and if the irradiation of the patient is kept down to the minimum consistent with the clinical needs of each case, the hazards are regarded as justifiable in the light of the benefits obtained.

The exposures to several children undergoing hip arthrography were measured by means of lithium fluoride thermoluminescent powder (Grech, 1972a). These radiation exposures ranged from 1500–2500 mR for cases using a combination of conventional radiography and screening, to 250–420 mR for 70 mm photofluorography and screening. It was found to be advantageous to have the screening current controlled automatically by the image brightness of the intensifier which therefore varied with the radiological thickness of the patient. With small children the current

Table 7   Radiation exposure in hip arthrography in children

Technique	Exposure (mR)	Exposure (mR) using lead screen filter
(a) Conventional radiography 65–70 kV; 16–20 mAs (5 films size 12·5×10 cm).	1350–2050	80–160
(b) 70 mm Photofluorography 65–70 kV; 100 mA (5 frames).	50–100	3–6
(c) Screening 55–65 kV 45 s – 2 min 20 s	100–1500	25–90

N.B. Exposures for (b) and (c) depend on patient thickness and 'preset dose rate' on the image intensifier.

indicated was usually below 0·2 mA at 55–65 kV. It was also found possible to reduce the patient's dose by placing a sheet of 18–22 gauge tinplate or lead intensifying screens on the couchtop beneath the patient. This additional filtration also had the effect of improving the T.V. image (particularly when using the video-tape recorder) by reducing the effect of 'shine-past' – the thick filter preferentially absorbing soft radiation. The reduction in the patient dose achieved by the use of a lead-screen filter can be seen in Table 7 (Column 2).

As it has been said before, in these hospitals this procedure is usually carried out on young children with a problematic dysplasic hip; consequently getting a definite diagnosis might prevent repeat plain radiographs.

In a survey in the practice of plain radiography in the diagnosis of hip dysplasia throughout the Sheffield region, with a population of approximately 4·5 million (assimilated in the Trent region since 1974), it was estimated that about 1000 new cases were referred every year for plain radiography with the provisional diagnosis of – or to exclude – hip dysplasia. It was also estimated that 44% of these children were under the age of six months (Grech, 1972b); we came across several cases where the

Table 8 Estimated incidence of CDH in children under 15 years in England and Wales 1962–72 inclusive.

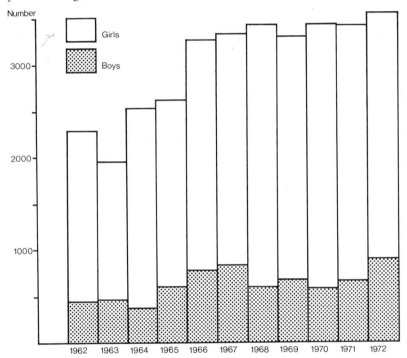

ESTIMATED INCIDENCE OF C.D.H. IN CHILDREN UNDER 15 YRS. IN ENGLAND AND WALES 1962–1972 INCLUSIVE

suspected hip was radiographed five times or more before these children were one year old. If this is extrapolated to cover the whole country, many children's hips are being radiographed for doubtful beneficial information.

The national incidence of 'congenital dislocation of the hip' in England and Wales in children under 15 years of age is given in Table 8. These figures have been extracted from 'Report on Hospital In-patient Enquiry' for the years reviewed, assuming that this was the final diagnosis on discharge of the patient. Unfortunately, it proved difficult to split these numbers into age groups. However, this table gives the impression that there is no appreciable decrease in the number of cases in this country. Actual increase in the incidence of hip dysplasia has been reported from some countries including Norway (Bjerkedal and Bakketeig, 1975); United States of America, (Center for Disease Control, 1975) and Israel, (Klingbert, M. A. *et al.*, 1976).

It might be worth reminding ourselves of the radiation dose received from conventional hip radiography – i.e. antero-posterior views of the hip using mainly circular cones and therefore irradiating a large field, including the gonads; according to Webster and Merrill (1957), using 66 kV 70 mAs, the skin dose was 790 mR and the doses received by the testes and ovary were 360 mR and 150 mR respectively. It might be relevant to draw our attention to the recommendation by the International Commission Radiological Protection (1970) which is the most authoritative statement available on the protection of the patient in X-ray diagnosis, that 'it is . . . not appropriate to give any predetermined, generally applicable dose limitations, and the ideal judgement of what should be considered to be an acceptable dose would have to be based upon known circumstances in each case, including the consideration of alternative techniques that may permit a dose reduction'.

In the light of this the level of radiation exposure from hip arthrography is not excessive and is acceptable, especially since, it is hoped, such an examination will give a definite diagnosis and obviates the need for repeated plain radiographs. The amount of radiation to such young children should be carefully watched, especially in such orthopaedic conditions where repeated X-rays are likely to be requested. Finally, it must be remembered that it is the duty of the radiologist to draw the attention of the clinician if the radiation dose is getting dangerously high.

References

Bjerkedal, T. and Bakketeig, L. S. (1975), Surveillance of congenital malformations and other conditions of the newborn. *International Journal of Epidemiology*, **4**, 31–36.

Center for Disease Control (1975), *Congenital Malformations Surveillance Report*, April 1974–March 1975.

Grech, P. (1972a), Video-arthrography in hip dysplasia. *Clinical Radiology*, **23**, 202–207.

Grech, P. (1972b), Arthrography in hip dysplasia in infants. *Radiology*, **38**, 172–179.

International Commission on Radiological Protection (1970), *Protection of the patient in the X-ray Department.* I.C.R.P. Publication No. 16., Pergamon Press, New York.

Klingberg, M. A., Chen, R., Chemke, J. and Levin, S. (1976), Rising rates of congenital dislocation of the hip. *Lancet*, **1**, 298.

Webster, E. R. and Merrill, O. E. (1957), Radiation Hazards – measurements of gonadal dose in radiographic examinations. *New England Journal of Medicine*, **257**, 811–819.

Index

Numbers in bold type represent the main reference